Uncertainty in Geographical Information

Research Monographs in
Geographic Information Systems
Edited by Peter Fisher and Jonathan Raper

Error Propagation in Environmental Modelling with GIS
G.B.M. Heuvelink (1998)

Causes and Consequences of Map Generalisation
E.M. João (1998)

Interoperable and Distributed Processing in GIS
A. Vckovski (1998)

An Introduction to the Theory of Spatial Object Modelling for GIS
M. Molenaar (1998)

Object-Oriented Modelling for Temporal GIS
M. Wachowicz (1999)

Spatial Multimedia and Virtual Reality
Edited by A. Câmara and J. Raper (1999)

Marine and Coastal Geographical Information Systems
Edited by D.J. Wright and D.J. Bartlett (1999)

Geographic Information Systems for Group Decision Making
P. Jankowski and T. Nyerges (2000)

Geographic Data Mining and Knowledge Discovery
Edited by H.J. Miller and J. Han (2001)

Uncertainty in Geographical Information
J. Zhang and M. Goodchild (2002)

Small-scale Map Projection Design
F. Canters (2002)

Uncertainty in Geographical Information

Jingxiong Zhang
and
Michael F. Goodchild

First published 2002 by Taylor & Francis
11 New Fetter Lane, London EC4P 4EE

Simultaneously published in the USA and Canada
by Taylor & Francis Inc,
29 West 35th Street, New York, NY 10001

Taylor & Francis is an imprint of the Taylor & Francis Group

© 2002 Jingxiong Zhang and Michael Goodchild

Produced from camera-ready copy supplied by the authors
Printed and bound in Great Britain by MPG Books Ltd, Bodmin

British Library Cataloguing in Publication Data
A catalogue record for this book is available from the British Library

Library of Congress Cataloging in Publication Data
Zhang, Jingxiong, 1964-
    Uncertainty in geographical information / Jingxiong Zhang and Michael
Goodchild.
        p. cm -- (Research monographs in GIS)
    Includes bibliographical references (p.)
    ISBN 0-415-24334-3 (alk. paper)
    1. Geographic information systems. 2. Uncertainty (Information theory) I.
Goodchild, Micahel F. II Title III. Research monographs in geographic
information systems

G70.212.Z53 2002
910.285—DC21

                                                                    2001053961

ISBN 0-415-27723-X

Contents

Series Introduction

Welcome

The *Research Monographs in Geographical Information Systems* series provides a publication outlet for research of the highest quality in GIS, which is longer than would normally be acceptable for publication in a journal. The series includes single- and multiple-author research monographs, often based upon PhD theses and the like, and special collections of thematic papers.

The need

We believe that there is a need, from the point of view of both readers (researchers and practitioners) and authors, for longer treatments of subjects related to GIS than are widely available currently. We feel that the value of much research is actually devalued by being broken up into separate articles for publication in journals. At the same time, we realise that many career decisions are based on publication records, and that peer review plays an important part in that process. Therefore a named editorial board supports the series, and advice is sought from them on all submissions.

Successful submissions will focus on a single theme of interest to the GIS community, and treat it in depth, giving full proofs, methodological procedures or code where appropriate to help the reader appreciate the utility of the work in the Monograph. No area of interest in GIS is excluded, although material should demonstrably advance thinking and understanding in spatial information science. Theoretical, technical and application-oriented approaches are all welcomed.

The medium

In the first instance the majority of Monographs will be in the form of a traditional text book, but, in a changing world of publishing, we actively encourage publication on CD-ROM, the placing of supporting material on web sites, or publication of programs and of data. No form of dissemination is discounted, and prospective authors are invited to suggest whatever primary form of publication and support material they think is appropriate.

The editorial board

The Monograph series is supported by an editorial board. Every monograph proposal is sent to all members of the board which includes Ralf Bill, António Câmara, Joseph Ferreira, Pip Forer, Andrew Frank, Gail Kucera, Peter van Oostrom, and Enrico Puppo. These people have been invited for their experience in the field, of monograph writing, and for their geographic and subject diversity. Members may also be involved later in the process with particular monographs.

Future submissions

Anyone who is interested in preparing a Research Monograph, should contact either of the editors. Advice on how to proceed will be available from them, and is treated on a case by case basis.

For now we hope that you find this, the 10th in the series, a worthwhile addition to your GIS bookshelf, and that you may be inspired to submit a proposal too.

Editors:

Professor Peter Fisher
Department of Geography
University of Leicester
Leicester
LE1 7RH
UK
Phone: +44 (0) 116 252 3839
Fax: +44 (0) 116 252 3854
Email: pff1@le.ac.uk

Professor Jonathan Raper
Department of Information Science
City University
Northampton Square
London EC1V 0HB
UK
Phone: +44 (0) 207 477 8415
Fax: +44 (0) 207 477 8584
Email: raper@soi.city.ac.uk

Preface

Geography is literally the representation of the Earth and, although today's discipline of geography has evolved substantially since the term was coined, representations remain a central issue to today's geographers. Representations of the Earth may take the form of maps, or images from space, or narrative text, or mathematical models. They may exist on paper, on photographic film, and increasingly in digital computers in the form of zeroes and ones. That last example provides perhaps the most striking instance of the fundamental paradox—digital representation of the Earth is about taking what we see around us, on the surface of this incredibly complex and beautiful planet, and reducing it to combinations of symbols in an alphabet with only two letters. Binary symbols must be used in digital computers whether the information that is being coded is about rivers, roads, terrain, or the dynamic behaviour of glaciers.

Representations inevitably simplify reality, and yet they appear to be faithful reproductions of reality. A child looking at a map is not immediately aware of all of the geographical information that is absent from the map, often because such missing information refers to details too small to be seen at the scale of the map, but also because the map reduces complex and poorly defined features like mountains or cities to simple shapes, and even to standard symbols. What the map does not reveal about the world is sometimes as important as what it does reveal, especially when important decisions have to be made based on maps.

By omitting details and smoothing irregularities, maps present clarified views of the world, and one can speculate that this is one reason humans admire and treasure maps—they simplify the world for us, and remove much of its awkward and confusing complexity. Digital maps, or maps stored in digital form in computers, extend this one step further, by presenting numerical data about the world, perhaps locations expressed in latitude and longitude, to great precision using many decimal places. The precision of many geographical information systems (GIS) is sufficient to locate independently every molecule on the Earth's surface. But of course we have no way of tracking every molecule, so such precision merely serves to provide a false sense of numerical security.

The inaccuracies and uncertainties of maps have interested scholars for decades, if not centuries. But the advent of computers, and GIS in particular, have given this interest new impetus and relevance. Maps are uncertain, but maps in vast quantities have been digitised and ingested into GIS, and used to make decisions, regulate activities, and conduct scientific research. Yet a map is clearly not a collection of scientific measurements in the normal sense of that term. Thus we face a mounting dilemma in GIS: clear-cut decisions are being made and enforced through regulations based on data that are known to their users to be uncertain and inaccurate. In such circumstances a court will clearly hold with a plaintiff who argues that the regulator failed to pay appropriate attention to uncertainty.

This book is about the research that has been conducted over the past two decades on the issue of GIS uncertainty. It presents results concerning the description and measurement of uncertainty in GIS data, its display in the form of maps, the modelling of uncertainty, the propagation of uncertainty through GIS operations and geo-processing, and its emergence in the form of uncertain decisions. The discussion draws heavily on the theoretical frameworks provided by spatial statistics and geostatistics, two areas that provide a rich source of theory and methods for tackling this inherently practical problem.

This line of research gained much of its early momentum from the first research initiative of the U.S. National Center for Geographic Information and Analysis (NCGIA), which was founded with funding from the National Science Foundation in 1988. NCGIA consists of a consortium of three universities—the University of California, Santa Barbara, the University at Buffalo, and the University of Maine. In December 1988 it held its first Specialist Meeting in Santa Barbara on the topic of Accuracy of Spatial Databases, bringing together some 45 scholars from around the world with interests in this problem. The meeting resulted in a book (Goodchild and Gopal, 1989) and led to a series of related initiatives. Several other books have appeared since then, as well as many journal articles, but no book has attempted to summarise the field in a single volume. We have written this book in the hope of meeting that goal.

Both the authors have been deeply involved in the development of this research field. Michael Goodchild began research on the application of concepts of geometric probability to GIS in the 1970s, he moved to the University of California, Santa Barbara, in 1988, led the first NCGIA initiative, and continues active research in the area. Jingxiong Zhang completed a PhD at the University of Edinburgh on uncertainty, and since early 2000 has been a postdoctoral researcher at NCGIA.

Many individuals have made significant contributions to research on uncertainty in GIS, and some were active in the field long before the NCGIA initiative. Jingxiong Zhang would like to acknowledge, in particular, the inspiration of Roger Kirby, Deren Li, and Neil Stuart. We would both like to acknowledge the pioneering work of Peter Burrough, Dan Griffith, Bob Haining, D. H. Maling, Benoît Mandelbrot, Paul Switzer, Waldo Tobler, and the many pioneers of the geostatistical school. Many other colleagues too numerous to mention have been sources of ideas over the years, but we would particularly like to thank Giuseppe Arbia, Peter Atkinson, Kate Beard, Nick Chrisman, Keith Clarke, Jane Drummond, Peter Fisher, Giles Foody, Gerard Heuvelink, Gary Hunter, Harri Kiiveri, Phaedon Kyriakidis, Kim Lowell, Ashton Shortridge, François Vauglin, and Mike Worboys. Rob Hootsmans, Gerard Heuvelink, and an anonymous reviewer provided helpful comments on the book proposal.

Finally, we thank our families for their support and understanding, without which the completion of the book would not have been possible.

Jingxiong Zhang and Michael F Goodchild
University of California, Santa Barbara
August, 2001

CHAPTER 1

Geographical Information and Uncertainty

1.1 GEOGRAPHICAL INFORMATION

Geographical (or geographic) information can be defined as information about features and phenomena located on or near the surface of the Earth (Goodchild *et al.*, 1999). Geographical information arises mainly from the practical need to solve geographical problems, some of which are indeed very pressing, being of major environmental and social concern, such as biological conservation, or of an immediate nature, such as finding a route for a trip. The gathering of geographical information is undertaken by government agencies such as the United States Geological Survey and the Ordnance Survey of Great Britain, and by various private sector companies.

This process of information gathering is traditionally performed by field scientists, and the results transcribed onto paper maps, which are traditional media and tools for geographical information and analysis. But advances in geographical information technologies such as remote sensing and geographical information systems (GISs, or GIS as a general term) have revolutionised the way geographical information is gathered and processed. Remote sensing of the Earth's surface from satellites has found many applications, such as land cover classification (Jensen, 1979; Campbell, 1996; Fuller *et al.*, 1998). GIS coupled with remote sensing are widely acclaimed as advanced technologies for managing and handling massive amounts of information relevant to geographical mapping, modelling, and understanding (Burrough and McDonnell, 1998). Developments in computing techniques have played an important role in redefining and expanding the data sources for mapping and geo-processing. Geographical information technologies have now reached an unprecedented position, offering a wide range of very powerful functions such as information retrieval and display, analytical tools, and decision support (for a general introduction to GIS and geographical information technologies in general see Clarke, 1995; Longley *et al.*, 2001).

Methods for GIS-supported problem-solving are not necessarily restricted to geography. Other disciplines such as agriculture, hydrology, landscape ecology, and soil conservation benefit from the use of geographical perspectives, and from the implementation of methodologies for addressing issues originating in disciplines other than geography. This is true particularly of digital methods for geographical information processing. A variety of applications have been made of GIS with varied success (Longley *et al.*, 1999).

GIS needs scientific orientation to sustain technological developments and operational values. A scientific orientation emphasizes accuracy in observations, the use of commonly accepted definitions to ensure accurate communication, and the need for two investigators to arrive independently at the same conclusions given the same information (replicability). However, proliferation of powerful GIS hardware and software does not necessarily guarantee such principles in the collection, processing, and use of geographical information. The scientific strength of GIS rests with fundamental research on geographical information, including concern for geographical aspects of cognition, measurement, computation, and decision-making, as discussed by Goodchild *et al.* (1999). This chapter, and indeed this book, will seek to enhance understanding of geographical information and contribute to its scientific coherence.

It is generally accepted that geographical information is distinguished from other types of information by its reference to specific geographical locations. A piece of geographical information may be reduced to a simple statement that there exists an instance of some more generally recognised thing at a specified location or place. The thing may be an entity, a spatial category, an activity, a measurement of some variable, or a concept. In a more general notion, these similarly recognised things may be termed geographically or spatially distributed phenomena. In brief, geographical information consists of facts or knowledge obtained from investigation, study, or instruction about specific places on the Earth's surface.

Spatial information, as a synonym of geographical information, seems to capture the spatial essence of geographies. However, there are genuine, although subtle, differences in the use of the terms geographical and spatial, despite the fact that both are often used almost interchangeably. Spatial information refers to information related to any multidimensional frame, which suggests a more general context than geography. For example, images taken of industrial assembly lines are analysed to provide information about the identification and location of abnormal parts posing quality concerns. Grenander (1996) proposed a theory of patterns, by which images of biological forms are analysed with mathematical constructs. Clearly, industrial and medical images and information are not limited to geographical space. On the other hand, geographical information is not necessarily limited to pieces of information with explicit reference to spatial locations. Geographical coordinates are even absent in geographical information in raster formats, such as satellite images, where locations may be implicit rather than explicit. Moreover, there may be little particularly special about a database that happens to contain geographical coordinates, as may be the case with fact-finders in a children's encyclopaedia that hardly qualify formally as GIS.

In general, geographical concepts formalised and implemented in GIS go far beyond point position, geometric concepts of direction and distance, or simple geometric primitives, to cover a much wider and more sophisticated set of geographical forms, patterns, and processes. These may include topological relationships of adjacency, connectivity, and intersection; concepts of simple and complex geographical objects from points, lines, and areas to watersheds; and the concepts embedded in models of processes that affect and modify the landscape. The elicitation of geographical concepts by humans contributes to developments in GIS, by helping to perfect the scientific tools used to represent, learn, and reason about, model, and understand the worlds that surround humans and shape societies and ecosystems.

Geographical information may be decomposed into primitive elements that take the form of tuples (x,G), similar to the geographical data matrices formalised by Chorley and Haggett (1968). In such itemised tuples, x refers to a general definition of location in two (X,Y) three $(X,Y,Z$ or $X,Y,T)$, or four dimensions (X,Y,Z,T), depending on whether the third spatial dimension and time are important or not, and G stands for one or more properties, attributes, or things, in other words selected geographical phenomena being studied at that location. If the thing is time-dependent, the location x should include time (T), and as well should refer to a certain spatiotemporal context. Thus, in its extended form, geographical information may be denoted as tuples (X,Y,Z,T,G), which means that some phenomena G are present at some location (X,Y,Z,T) in space-time.

These tuples are commonly termed geographical data, and consist of data referenced spatially using such locators as plane coordinates or geographical coordinates (Maling, 1989). For example, a town in a coarse-scale map may be usefully represented as a point, with coordinates to define its position, and can be attached with facts known as attribute data, such as the name and population of the town. To some extent, data stand as facts used for calculation and reasoning, and are the carrier and media for information processing, while information seems to be more abstract and is thus discussed in a more general sense. Similar to the relationship between geographical and spatial information, geographical and spatial data may be used as synonyms. The newer term *geospatial* connotes geographical as a special form of spatial, and is effectively synonymous with geographical in practice.

Remote sensing plays an important role in the provision of geographical data for GIS and environmental modelling, of which land cover and biomass mapping are operational applications. Photogrammetry is also a mainstream technique for data production, and impressive recent developments in this field are directed towards automation by digital or softcopy photogrammetry (Mikhail *et al.*, 2001). In digital photogrammetry, digital images are input into a computer, and semi-automated techniques are used to derive a variety of spatial data sets, including digital elevation models (DEMs) and digital ortho-photo quadrangles (DOQs), leading to greatly enhanced productivity.

Given that the information inputs are accurate, up-to-date, and complete for a specific problem domain, the substantial computational sophistication of mainstream GIS implies that output information, or solutions obtained using the technology, will have levels of accuracy that are numerically reasonable and comparable to those in the input data. This is a picture that has been encouraged by the digital nature of GIS, and GIS's legacy of cartography.

1.2 UNCERTAINTY

Unfortunately, geographical data are often analysed and communicated amid largely non-negligible uncertainty. Error-laden data, used without consideration of their intrinsic uncertainty, are highly likely to lead to information of dubious value. Uncertainty exists in the whole process from geographical abstraction, data acquisition, and geo-processing to the use of data, as the following text develops.

The geographical world is infinitely complex, being the result of the distribution of almost 100 distinct atomic species, at densities that range from order 10^{23} molecules per cubic centimetre. Magnetic, electrostatic, and gravitational fields vary continuously at multiple scales, as do such abstracted properties as temperature, pressure, pH, and fluid velocity. Geographical reality, though immensely complex, needs to be necessarily abstracted, simplified and discretised to facilitate analysis and decision-making, in particular, when using digital methods. In a digital environment, approximation must be expressed in a limited number of digits, as computer storage space is always limited.

Geographical abstraction is selective, entails structuring of the web of reality, and leads to approximation, as noted by Chorley and Haggett (1968). The processes of abstracting and generalising real forms of geographical variation in order to express them in a discrete digital store are defined as data modelling (Goodchild, 1989), and this process produces a conceptual model of the real world. It is highly unlikely that geographical complexity can be reduced to models with perfect accuracy. Thus the inevitable discrepancy between the modelled and real worlds constitutes inaccuracy and uncertainty, which may turn spatial decision-making sour.

Modelling represents a complex interaction of human and instrumental factors, and acquiring the raw components of each model, the data themselves, is also subject to a host of uncertainties. Depending on the skill of data analysts and the sophistication of instruments, data acquisition will have varying levels of error. Though advances in field technologies, such as global positioning systems (GPSs), and improvements in laboratory techniques, such as digital photogrammetry and digital image processing, have greatly improved the quality of spatial data, it is simply not possible to eliminate errors in data acquisition, and this is true of measurement generally.

Whether through sensing of spatial phenomena using automated or remote means, or despite the high costs incurred in collecting data on the ground by direct human observation, there is a growing quantity of data that may be used to cater for a specific purpose. As there are often many sources of data representing the same phenomena, it is common to encounter different GIS software systems of different standards, formats, terminology, and practice. In the absence of interoperability between major GIS software products and data formats, it can be very time-consuming to have to transfer data from one system to another (Vckovski, 1998). Errors occur during such geo-processing. A good example is the conversion of a vector data structure to a raster data structure or vice versa: the converted data rarely duplicate the original data, even if the original data are error-free, leading to so-called processing errors.

With more and more sources of data, it is becoming increasingly likely that heterogeneous data of varying lineage and accuracies will be involved in driving a query or arriving at a choice. Methods for combining data sets will have to cope with and resolve spatial and semantic differences between heterogeneous data sets, and will have to derive the best possible data from them. Hutchinson (1982) investigated methods for improving digital classification of remotely sensed images by incorporating ancillary data. Conflation of heterogeneous spatial data remains an issue of theoretical and practical importance (Saalfeld, 1988), in which uncertainty will inevitably play a crucial role. However, methods for integrating spatial data may not be comparable to the simple concept of a weighted average applied to non-spatial data, whose statistical treatment is well documented and generally straightforward (Goodchild *et al.*, 1995). This is because spatial data have complex, multi-dimensional properties that require special handling and processing, as shown by Atkinson and Tate (2000).

This is a more apparent concern in the case of urban applications where many complex spatial entities are often superimposed and compressed into small areas. Accordingly, there is a need for fine-scale map data on the one hand, such as maps recording buildings with high precision, and less detailed data at coarse scales on the other hand, such as highly generalised coverages of population density. Such a situation will inevitably lead to great complexity in error propagation, when data layers of varying accuracies and levels of detail are combined, that is, when data are integrated to derive certain desirable products or to assemble the raw data required for spatial analysis or modelling.

As geographical data have their inherent (and often limited) accuracies and different levels of detail, data combination must adapt to spatially varying accuracies in component data sources if it is to be scientifically sound. But the usual data integration procedures take no account of the varying levels of accuracy and detail of the data sets being merged, and instead assume scaleless and precise digital data during geo-processing (Abler, 1987). As a result, when digital maps of different scales—for example, a detailed and accurate city map and a highly generalised stream network map—are merged or overlaid by adopting a common scale, the results may be meaningless or potentially dangerous. For example, streams from a highly generalised map may be significantly out of position when compared to features on a much more detailed map. This can explain the occurrence of error propagation, or more specifically error concatenation, during such geo-processing as map overlay, if source data are not handled with awareness and accommodation of spatially varying and maybe correlated errors.

Therefore, with geographical models and measurements suffering approximation and inaccuracy, geographical information can no longer be considered as error-free tuples recorded in spatial databases. Hypothetical true tuples (x, G) may be more objectively written as their measured or, more generically, realised versions (x', G'), with differences denoted by $x–x'$ and $G–G'$ (note that this difference between true and observed properties is appropriate only when such properties are measured on interval or ratio scales). Alternatively, values of (x', G') may be associated with probability distributions, to represent their uncertainties.

Data may pass through many different transactions and custodians, each providing their own protocols or interpretations. Thus, uncertainty is not a property of the data contents so much as a function of the relationships between data and users. In this context uncertainty may be defined as a measure of the difference between the data and the meaning attached to the data by the current user. In other words, uncertainty is a measure of the difference between the actual contents of a database, and the contents that the current user would have created by direct and perfectly accurate observation of reality. In an alphanumerical world, there is no inherent check against potential misuse of digital geographical data. In particular, detailed information about primary sources of data, which may be acquired from field or photogrammetric surveys, or about data derived from secondary sources, is generally not available to potential users. Most users must rely on published maps or data products provided by government agencies such as the U.S. Geological Survey, and must take such products at face value. So, there is an increased need for information on uncertainty in a GIS-based environment to assess the fitness-for-use of data.

Traditionally, errors are annoying and to be avoided or reduced to minimal levels. In the previous discussion, uncertainty and error seem to have been used almost interchangeably. Errors refer conventionally to inaccuracy and imprecision, and are used with an implicit assumption that true values are obtainable, at least in principle. It is easy to arrive at the conclusion that errors are computable if truth is obtainable, or truth is recoverable if errors become known. However, geographical data recorded in GIS are actually representations of the underlying phenomena, that is, tuples (x',G') rather than (x,G). In many situations, it would be extremely difficult, if not impossible, to recover the truth fully, and in many situations it is difficult to believe that a truth actually exists. For example, it may be impossible to define a true soil class if definitions of soil classes are inherently uncertain or vague, as many are. Thus, the notion of uncertainty can be a more realistic view of geographical measurement than the notion of errors, which suggests attainable true values. A re-oriented view of GIS-related errors is that they should be conceived as an integral part of human knowledge and understanding concerning geographical reality, so that information is provided and used with well-informed assessments of uncertainty. In this regard, uncertainty captures the information richness underneath the nominal contents of data sets, and measures the degree to which the database leaves its users uncertain about the nature of the real world.

Statistical estimation and inference, which are extensively used with non-spatial data, usually assume homogeneity and independence with regard to a specific population. However, this gets problematic when dealing with the heterogeneity and dependence that are intrinsic to many geographical data and, of course, much geographical information and analysis. Moran (1948) examined the interdependence of geographical distributions labelled on statistical maps, and derived moment quantities for assessing whether the presence of some characteristics in a location (county) makes their presence in neighbouring counties more or less likely. A test by Campbell (1981) using Landsat MSS (multispectral scanner) data for rural land cover classification revealed the presence of positive continuity between reflectance values in adjacent pixels, which led to biased estimates of category variances. These biased estimates led further to misclassification, and erroneously classified pixels tended to cluster. Many urban fabrics and man-made structures imaged in aerial photographs are readily identified visually, and can be mapped more accurately than rural landscapes. Even within a single image scene, locations closer to ground controls will have more accuracy in positioning than those further away. Moreover, the statistics of joint distributions of two spatial neighbours will be different depending on whether their occurrences are correlated or not (Goodchild *et al.*, 1992).

Last but not least, ambiguity, inexactness, and vagueness exist in many geographical studies and applications, reflecting the inherent complexity of the geographical world and the subjectivity involved in a high proportion of geographical data (Black, 1937; Ahlqvist *et al.*, 2000). The concept of agricultural suitability of a land holding has an inherent lack of exactness. Urban heat islands are vaguely defined spatially, and their hazardous status is also vaguely communicated even with quantitative indices. In such domains there will be hardly any true values that are comparable to those of precisely defined variables such as land ownership and taxable holdings. Use of the term uncertainty seems to signal an appreciation of vagueness as well as randomness in geographical information, while error may send a misleading signal that true values are definable and retrievable. Besides, uncertainty may be a result of lack of information, and the amount of uncertainty is related to the quantity of information needed for recovering the truth. It is thus argued that uncertainty seems a more relevant concept than error, in particular for geographical information, which is characterised by heterogeneity, spatial dependence, and vagueness. Therefore, the term uncertainty will be used in this book as an umbrella term for errors, randomness, and vagueness, although uncertainty and error are used as synonyms at times where the context is unlikely to cause any confusion.

1.3 RESEARCH TOPICS AND SCOPE OF THE BOOK

The goals of research on uncertainty in geographical data, according to the University Consortium for Geographic Information Science (UCGIS) (1996), are to investigate how uncertainties arise, or are created and propagated in the GIS information process; and what effects these uncertainties might have on the results of subsequent decision-making. If it is possible to get information on uncertainties intrinsic to geographical data during geo-processing, and to track the propagation of these uncertainties, then both producers and users can assess the accuracy obtainable from the analysis of a certain composite map (Goodchild, 1991). Systematic and rigorous work on uncertainty modelling is required to deal effectively with the inevitable impacts of uncertainty on decision-making. This book aims to contribute to this process.

At the root of the problem is the need for effective identification of the sources of uncertainty. This is understandable, since errors in data sources incorporated in a specific geographical application are likely to have the most profound impacts on the resultant information products and decision-making. Thus, special attention must be given to the discussion of sources of uncertainty in geographical information and analysis.

As has already been mentioned in the previous section, different sources of uncertainty interact to affect geographical information and its analysis (Chrisman, 1991). Uncertainty may be of conceptual nature, and associated with the process of abstraction or generalisation about real world phenomena. As an example of conceptual uncertainty, there may be a difficult decision to make about whether a small patch of trees in a park, which differs from the surrounding region of grass, should be picked up as a separate class, or blended with grass into the large entity of parkland. In other words, are trees a recognised component of the park class, or a separate class in their own right? If spatial variability and heterogeneity are important characteristics for class mapping, the trees should clearly be retained and considered as part of the geographical distribution of categories in a larger context such as land classification. On the other hand, for land use management, the trees are hardly significant enough to be mapped, if they are of small areal extent and little historical or cultural importance.

Uncertainty may occur due to errors in measuring the positions and attributes of geographical entities, or in sampling geographically distributed phenomena. There are many occasions where measurement errors occur. A data operator can hardly track precisely the lines to be digitised during a map digitising task—the light dot is not accurately kept to the ground in sampling elevation data on a photogrammetric plotter; an agricultural field is wrongly classified as a recreational area in a classified remote sensing image.

Inappropriate processing of source data may be prone to error, even when the data are themselves accurate enough. For geo-processing, Goodchild and Lam (1980) described errors in the spatial interpolation of area data. Aronoff (1993) provides a good summary of the common sources of errors encountered in GIS applications, which shows how errors occur from the data sources, through data storage and manipulation, to data output and use of GIS data results.

The next issue is to detect and measure uncertainty, that is, to assess accuracy levels in geographical data. Many techniques for accuracy assessment developed in land surveying, and in the mapping sciences in a broader sense, are useful and may be adapted for uncertainty handling in GIS. Error ellipses for points and epsilon error bands for lines are widely used to describe and visualise positional uncertainty in point and line entities stored in spatial databases (Dutton, 1992). Classification accuracy in the remote sensing of landscapes is commonly evaluated by measures such as the percentage of correctly classified pixels, which have been adapted from statistics of non-spatial data measured on nominal scales (Cohen, 1960; Hord and Brooner, 1976).

Following identification and evaluation of uncertainty, there is the issue of uncertainty propagation when heterogeneous spatial data are integrated for spatial problem-solving. An analysis of uncertainty propagation takes information about uncertainty in the data inputs to an analysis, combined with knowledge of the analysis process, and makes predictions about the uncertainties associated with the outputs. Thus, it is possible in this way to predict the consequences of using a particular set of data for a specific purpose. To some extent, uncertainties are better modelled as if they were data about spatial data. Thus, a general approach to uncertainty modelling is to devise techniques by which equal-probable realisations of geographical data conforming to inherent distributions and characteristics can be generated, and used to examine the range of possible outputs.

Error ellipses based on Gaussian distributions provide such models, as well as descriptors for positional uncertainties in points (Wolf and Ghilani, 1997). Goodchild *et al.* (1992) described a model for uncertainties in categorical maps, which uses probabilistic maps and a spatial autocorrelation parameter together to characterise generic problems of heterogeneity within polygons, and transitions across polygon boundaries. These probabilistic maps can be derived from an image classifier, from the legend of a soil map, or through subjective assessment of uncertainties of interpretations. Goodchild (1989) defined some of the basic concepts of uncertainty modelling, and Heuvelink (1998) reviewed the increasing investment in conducting and publishing research on error propagation over the past decade.

The final issue refers to strategies for error management, that is, decision-making in the presence of errors, and strategies for the reduction or elimination of error in output products. These two interconnected problems, being of equal importance, go beyond error assessment and are concerned with the inferences that may be drawn from the results of error propagation. A detailed discussion is well beyond the scope of this book.

In summary, research on uncertainty is concerned with uncertainty identification, description and modelling, visualisation, and with predicting its effects on the outputs of analysis given knowledge of uncertainty in inputs. While substantial progress has been made since the book by Goodchild and Gopal (1989) was published, there is a need to synthesise past work on geographical data and uncertainties, and to build a coherent framework of theories and methods with the goal of increasing the availability of information about geographical uncertainties.

It is easy to see from the hierarchy of uncertainty issues described above and suggested by Veregin (1989) that uncertainty modelling stands at a level higher than identification and description of uncertainties, and is the key to uncertainty propagation. In other words, modelling is the core of uncertainty handling, since spatial modelling of uncertainties offers the potential for fuller use of uncertain geographical information, going far beyond what is possible with descriptive statistics of spatial data, let alone simple maps.

The search for models of uncertainty must start with an understanding of how geographies are abstracted and geographical data are modelled. There are two alternative perspectives on geographical abstraction, which, as mentioned previously in the example of deciding whether to map the trees out of the grassed terrain surrounding them, create a conceptual model of the real world. These two perspectives of geographical abstraction are called field-based and object-based, depending on whether geographical reality is regarded as consisting of a set of single-valued functions defined everywhere, or as occupied by a collection of discrete objects (Goodchild, 1989).

Different perspectives on geographical abstraction create different data models. Field-based models are favoured in physical geography and environmental applications, while object-based models are more suitable for cartography and facilities management. The choice between object-based and field-based models depends on the specific nature of the underlying phenomenon, and is limited by the data acquisition techniques and implementations of geographical databases, and will affect the possibilities for uncertainty modelling (Goodchild, 1993). Detailed accounts of object-based and field-based data models will be provided in later chapters. But it is clearly sensible to formulate object-based geographical information through the enumeration of objects as tuples (x,E) meaning an object E is georeferenced by x and may be characterised with certain attributes, while field-based geographical information can be put simply as functions $G(x)$ of a location. In two-dimensional geographies, $G(x)$ may be written as $Z(x)$ by viewing G as projected in the vertical dimension Z.

Uncertainties are better modelled in fields rather than in objects, because a field perspective facilitates the integration of the roles of data acquisition and analysis, thus allowing for the raw data with their spatial heterogeneities to be retained before compiling a specific map. These raw data are usually vital for deriving necessary information on errors in geo-processing (Goodchild, 1988; similar observations have been made by Bailey, 1988). Moreover, uncertainties in objects can then be seen as outcomes or realisations of a stochastic model defined on fields, extending a stochastic paradigm into the domain of discrete objects (Goodchild, 1993). Visualisation helps to enhance communication of field and object uncertainties in geographical analysis and decision-making, leading to improved knowledge about the underlying reality. As metadata, uncertainties may exhibit certain spatial patterns beyond pure randomness, which can be visualised and used to advantage (Goodchild, 1998). The advantages of field-based models are particularly significant during the combined use of different data layers that possess different levels of accuracy. Lastly, the field-based models provide an open strategy for research on error issues in GIS, especially when one recognises the potential of geostatistics (Journel, 1996).

With these considerations, this book pursues GIS error issues from a field perspective, though object-based models and methods are discussed for discrete objects, as reflected in the organisation of the book's chapters. The book discusses theoretical and practical aspects of spatial data and uncertainties with emphasis on the description and modelling of uncertainties within a systematic framework.

The first chapter has introduced concepts related to GIS-based spatial problem-solving, and the assumptions implied in GIS-based spatial analysis and modelling. This has been followed by an exposure of the consequences of imposing unsuitable models and assumptions and using uncertain data. Issues of uncertainty have been discussed with respect to research objectives in general terms. An overview of the book chapters concludes the introductory chapter, and provides an outline for the structure and organisation of the book.

Chapter 2 discusses fields and objects, the two models for geographical abstraction, with respect to how uncertainties in fields and objects occur. Fields are concepts of scientific significance, which have been applied widely in GIS with powerful analytical capability. Field variables may be continuous or categorical. Fields permit flexible mathematical analysis, for which slope and aspect are given as examples of continuous variables, while set operations such as AND and OR are discussed with hypothetical examples in land suitability assessment. Spatial variability is a fundamental property of fields, while spatial dependence provides further clues to a fundamental understanding of fields. For quantification of spatial dependence, a description is given of geostatistics, by which spatial variability and dependence are modelled with semivariograms and covariances. With the availability of a set of samples, complete field models may be constructed by exploiting spatial dependence in the process of spatial interpolation. On the other hand, objects are considered with respect to their formalism, and with respect to geometric and topological analysis. An object has position and a set of attributes. In the object domain, spatial coordinates are used to define locations or boundaries for objects that are identified as crisp sets of points, lines, and areas. Attributes are associated with an object as a whole. Objects are linked with certain types of topological relationship. Simple objects are agglomerated into complex objects by means of object-oriented approaches. A dialectical discussion of fields and objects is provided. Objects may be derived from fields, and the creation of hydrological features, such as drainage networks from DEMs, provides a good example.

Chapter 3 looks into geographical sampling (measurement) and various data commonly incorporated in GIS and spatial applications. It views uncertainty occurrences as the combined effects of information technologies and the partly subjective decisions of personnel employed for geographical sampling and data collection. Geographical measurement may be performed through ground-based surveys, or based on remote sensors. Land surveying remains the most traditional yet crucial method for geographical data collection, as it involves direct contact between human observers and reality. Nevertheless, remote sensing techniques have provided greater efficiency for spatial data collection. Depending on the specific sensor data formats, computerised methods drawing upon digital image analysis may be exploited, while visual interpretation and manual delineation constitute the main components of analogue or analytical photogrammetric approaches to mapping. Classification algorithms transform continuous variables to categorical ones as in remote sensing, while point categorical samples can be extended to areas as in the case of photo-interpretation. Certain continuous variables may be derived from remote sensors; examples include elevation, and assorted properties of vegetation. Map digitising remains a common method for extracting digital spatial objects from paper maps. Map digitising by manual tracking is often fraught with human fatigue and errors. Automation in map data collection must rely on human expertise for overcoming technical difficulties in object extraction from scanned maps in raster format.

Chapter 4 serves to set the context for uncertainty modelling in the chapters to follow. Methods for uncertainty handling have evolved from accuracy-oriented to spatially exploratory approaches. In accuracy-oriented approaches, the focus is placed on positional and attribute accuracy, lineage, logical consistency, and completeness, representing spatial extensions of statistical concepts and methods such as probability, mean, standard deviation, and other metadata. Uncertainties are discussed in line with data types including points, lines, areas, and surfaces, leading to a largely descriptive view for analysis and representation of highly complex spatial phenomena. For spatially exploratory analysis, on the other hand, uncertainties are viewed as spatially varied but correlated phenomena. This view fits neatly with originally continuous fields and is also adaptable for discrete spatial entities, although error fields are localised at object boundaries in the latter case. Another important feature of this chapter lies in the use of both probabilistic and fuzzy frameworks for uncertainties in both continuous and categorical variables, and in positions and attributes of objects.

Chapter 5 is concerned with uncertainties in continuous variables. Concepts such as image resolution and aggregation levels implicitly set the upper limits for accuracy in fields. Traditional measures for uncertainty, such as root mean squared error (RMSE), provide global estimates of errors, but their use is severely limited. Kriging produces a field of spatially varying variance along with an estimated field of the variable under consideration, whose probabilistic evaluation relies, however, on the assumption of normality in data distribution. Moreover, Kriging variance alone would not be able to allow one to evaluate the impacts of uncertainty in source data upon geographical inquires and analysis, since uncertainty propagation in a geographical context hinges on quantification of both data inaccuracies and spatial dependence therein. In other words, for propagating uncertainty from inputs to outputs, one would require modelling uncertainty, an endeavour more important and, of course, challenging than providing a description about uncertainty. Though analytical approaches such as variance propagation may be used for smooth and differentiable functions of field variables under simplified circumstances, their use is undermined by the spatial correlation that is central to many spatial problems involving manipulation of single or multiple variables. Stochastic simulation is more versatile and is widely applied for generating equal-probable realisations of a variable, which can, in turn, be used to drive modelling of uncertainties in complex geo-processing. Geostatistics and spatial statistical analysis are discussed for designing stochastic simulators. Examples concerning elevation and slope are given to illustrate the distinctions between analytic and simulation approaches to modelling uncertainties. Both manually sampled elevation data and semi-automatically generated DEM data are tested with geostatistical as well as more conventional approaches.

Chapter 6 is about spatial categories and various aspects of uncertainties regarding categorical mapping and the occurrences of categories with reference to land cover mapping. Accuracy measures such as percent correctly classified (PCC) for categorical variables provide little information on spatial dependence of categorical occurrence and have limited use for probabilistic modelling of spatial categories. Probabilistic mapping is performed by indicator Kriging on the bases of both graphical and digital images. Different methods for combining probabilistic categorical data are documented. Stochastic simulation of spatial categories is then discussed and compared with Kriging in terms of the expectation and variance of categorical occurrences. Since vagueness or fuzziness is another aspect of categorical uncertainties inherent in any process involving qualitative reasoning, fuzzy sets are described and compared with probabilistic methods for representing and handling uncertainties in categorical variables using real data sets. This comparative study of probabilistic versus fuzzy methods for categorical uncertainty handling serves to clarify their distinctions in a convincing way. Rough sets follow the discussion of probability and vagueness in spatial categories, and their potential use is highlighted with respect to the indiscernibility of data classes.

Uncertainties in objects are the theme for Chapter 7. As in the case for fields, discretisation in the object domain and related issues are discussed. This is followed by a description of uncertainties in objects, which consist of their positional and attribute dimensions, and above all, uncertainty in object abstraction and extraction. It is asserted that the emphasis for uncertainties in objects should be placed on position. Error ellipses and epsilon error bands, as descriptors for points and lines respectively, are examined with vector data sets. Modelling of uncertainties is then examined in the domain of objects where uncertainties in lines and areas are analysed with respect to lengths (perimeters) and area extents via fields of positional errors.

Chapter 8, the final chapter, gives some retrospective and prospective views. In looking back over the book, it is clear that the emphasis in uncertainty handling has been upon fields and objects, while the concepts and methods explored in the book have been drawn from geostatistics, probability theory, fuzzy sets, and rough sets. This retrospective is followed by a discussion of what the future holds for GIS-related communities that care about uncertainties; and of prospects for further developments.

In overview, there are specific novel contributions in this book, which are briefly discussed here. Much work on using field models and methods for representing and handling uncertainties is oriented towards the integration of environmental modelling with GIS. In the context of urban and suburban mapping, where spatial data of various sources, different accuracies, and diverse formats are usually involved, systematic approaches are seen to be lacking. This book endeavours to bridge this gap, and to construct integrated approaches to handling the uncertainties encountered in both built and natural geographies, where discrete objects and continuous fields are complementary conceptualisations of the world. In the interests of greater theoretical strength and practical applicability, this book synthesises past work pertinent to both the object-based and field-based models of uncertainties, and expresses it in a rigorous way.

Secondly, this book reassesses various theories and methods for uncertainty reasoning, including frequentist and subjective probabilities, fuzzy sets, and rough sets, to provide a complete treatment of different forms of uncertainty. As hinted previously, although probability has a longer history and sounder foundation than younger and less developed fuzzy sets and rough sets, the latter are surely relevant and valuable since subjectivity is an indispensable element in discretising and processing geographical information. While the empirical, or frequentist, concept of probability lends itself to relatively straightforward quantification, the subjective interpretation of probability may not be uniquely determined, because the way it represents a relation between a statement and a body of evidence is not purely logical but reflects degrees of personal belief. Prior knowledge in the form of prior probability is required in Bayesian inference, a statistical method in widespread use. However, obtaining prior information is not easy but requires collecting large quantities of empirical data, a substantial drawback of applying probabilistic protocols. Many geographical phenomena are characterised by heterogeneity in geographical and semantic terms, and their cognition and conceptualisation entail vagueness. Approximation is yet another necessary element in geographical information, since it is rarely possible and practical to possess perfect knowledge about reality. What usually works is to extract useful understanding of the problem domain by exploiting accumulated knowledge in the hope of reducing imprecision, given what is known of the domain. In practical settings, one is likely to encounter different forms of uncertainty, due to random variation, judgmental factors, indeterminate borderline effects, or a general lack of imperfect information, suggesting that a single reasoning mechanism, either probabilistic or fuzzy, is rarely adequate to model the complexity of uncertainty. One should proceed by using several approaches in combination to address uncertainty. For instance, the theoretical framework of probability may be relaxed to accommodate fuzziness in observations, and higher-order imprecision in fuzzy sets may be addressed by probability. This book attempts to bring together various methods for reasoning about uncertainty, and to capture their distinctiveness and complementarity.

The third characteristic of this book pertains to exploring geostatistical methods, not only for spatially aware treatment of uncertain geographical data, but also for conflating heterogeneous data. Applications of geostatistics for these purposes exploit information mechanisms underlying data sets characterised with varying inaccuracies and granularity. As spatial variability and dependence are intrinsic properties of many spatial distributions, their incorporation should be advantageous in processing and communicating geographical information. Spatial correlation can be usefully accommodated in both field and object domains, and so can cross-variable correlation, as this book explains. It will also be shown that geostatistics offers great potential for the consistent and theoretically rigorous weighting of data sets during conflation. It will become apparent that geostatistics is enabling, in that it adds value to geographical information by exploring uncertainty in depth, thus leading advanced and viable developments in geographical information and processing.

Lastly, conceptual, theoretical, and methodological frameworks, however elegant, need to be backed up by practical considerations concerning implementation. Towards this end, this book describes experiments with real data sets pertaining to both fields and objects, which were acquired first hand in close-to-real-world settings. These data were analysed through geostatistics, supported by the public-domain software system GSLIB, allowing uncertainty to be approached through description and stochastic simulation.

Uncertainties are something that will always pervade geographical information and its handling as a necessary and inescapable phenomenon. They are specific to the underlying data types, and will be dependent on application objectives. Ideally, uncertainties are catalogued and packaged neatly with the data sets, and methods (e.g., computer subroutines) are devised for assessing their fitness of use in GIS applications. However, before any standards and policies with respect to uncertainty are implemented, a feasible way forward is for data producers to be uncertainty-aware and conscientious so that spatial query and decision-making are uncertainty-informed with respect to heterogeneous data sets and specific applications.

Uncertainties in geographical information and analysis may be conceptualised within a scientific framework. An implication of this understanding is that recognising that uncertainty and errors are pervasive in geographical information is one significant step towards redefining GIS as science (Wright *et al.*, 1997). GIS implies algorithms, data models, and other elements of information technology that are necessary for the robust and efficient handling of geographical information. But most importantly, GIS should be considered as a coherent body of principles, concepts, and theories that are crucial for the promotion and advancement of geographical information science (GIScience; Goodchild, 1992). To strengthen and reinforce the scientific aspects of GIS, a variety of research topics have been identified and addressed as fundamental themes of GIScience (Abler, 1987; UCGIS, 1996). It is hoped that this book will advance GIScience through its examination of one of the most prominent and pervasive characteristics of geographical information—its many forms of uncertainty.

CHAPTER 2

Geographical Perspectives

2.1 INTRODUCTION

Well before the turn of the new millennium, GIS communities in the broadest sense had begun building national spatial data infrastructures and coordinated digital representations of the Earth (Digital Earth), and researching their scientific, technological, industrial, and social impacts (Coppock and Rhind, 1991; Mapping Science Committee, 1993; Goodchild, 1999b). Unlike the earlier debates over vector- or raster-based data structuring, when computers were used for simple operations such as displaying and analysing digitised maps (Peuquet, 1984, 1988), these efforts represented a much more comprehensive need to address the many roles of GIS and geographical information in complex modern societies. More fundamental still is the philosophical and scientific evolution of GIS, for which issues such as data structures and numerical algorithms are seen as primarily technical and limited in scope.

A prerequisite for GIS-based geographical studies and applications is the creation of models of the geographical world that respect both scientific and cognitive perspectives. In order to model the real world, it is necessary to abstract and generalise. Because most spatially distributed phenomena exist as complex and multivariate processes, varying at many scales in space and time, geographical, environmental, and Earth scientists have to select the most important aspects of any given phenomena and to use these as the basis for information abstraction, extraction, representation, and communication.

The process of abstracting reality is related to geographical conceptualisation, which determines how geographies are modelled and how geographical data are represented. Data models are selected to facilitate geographical analysis and problem-solving, although data models are sometimes technological impositions. Objects and fields are the two differentiated yet closely related types of data models for geographies. For example, in tourist information centres, geographical information is typically represented in the form of mileage links between places denoted as points, by which visitors may be able to travel to their desired destinations. Computerised transportation systems may well consist of databases of roads and streets, giving information about addresses and routes (Noronha *et al.*, 1999). These same places may be represented as polygonal objects that have been explicitly defined by boundary lines for other purposes such as demographic mapping (Bracken and Martin, 1989). In dealing with environmental problems, on the other hand, the concentration of pollutants in an area tends to be represented as contours or grids (raster), which are perhaps the most commonly used field models. Air-borne or space-borne remote sensors supply spatial data increasingly in digital raster forms, leading to a preference for gridded field models for data handling.

Uncertainties have been described initially in the introductory chapter as an integral part of geographical information and, in addition, of the underlying phenomena of interest. It is logical to explore uncertainties in the light of how geographical reality is modelled, measured, represented, and processed in the context of GIS-supported spatial problem-solving. As the process of geographical abstraction or data modelling generally occurs before other information processes, this chapter seeks to establish the field and object conceptualisations of geographical phenomena as the foundation for a discussion of geographical information and many aspects of uncertainty.

Object data models were developed in the past largely for the purpose of digitising the contents of paper maps. Employing discrete points, lines, and areas to represent spatial data seems to be compatible with the typology of real world entities, whose positions and attributes may be acquired by direct or remote measurement, as is done by surveying and mapping specialists. For example, telecommunication cable networks laid underground or in deep oceans can be represented via discrete points and lines in digital maps to facilitate precise recording and efficient management of the networks. Geo-relational representation schemes are an obvious option, and handling such data sets seems a familiar task to the computer-minded.

Objects are typically connected or contiguous, as in the examples of buildings and land parcels, but sometimes scattered or separated, as is the case for cartographic rendering of surging weather fronts. Objects are identified primarily by their non-spatial characteristics, such as land tenure and property ownership, and then demarcated with their spatial (and temporal) extents. Thus, under a discrete object-based model, spatial data represent geographical reality in the form of static geometric objects and associated attributes. Accordingly, a GIS built upon discrete objects is largely limited to primitive geometric operations that create new objects or compute relationships between objects.

Despite the influential cartographic tradition, and the popular use of discrete objects for everyday purposes, many map features are the products of geographical abstraction, or technical consequences of the constraints of the map medium, and do not exist in the real world. For instance, boundaries of soil types that are neatly drawn less than 0.5 mm wide are actually transitional zones between neighbouring polygons. Contours are not marked on the terrain for humans to find. There are also artificial objects created for the purposes of the database, for example pixels in raster images. It is common knowledge that lakes fluctuate widely in area, yet have permanently recorded shorelines on maps. Therefore, it is difficult to use object models alone to represent multi-resolution, dynamically changing geographies on the curved surface of the Earth, because what can be represented by discrete and exact-looking objects on a paper medium with a single and uniform level of resolution is extremely limited.

Many geographical phenomena exist everywhere and vary continuously over the Earth's surface. Elevation of terrain, atmospheric temperature and pressure, vegetation and soil types are common examples. As these phenomena are continuous over space (and time), environmental models designed to describe the processes that operate on and modify the landscape are often in the form of partial differential equations, which implicitly recognise the continuity of space and the constantly changing values of the independent variables. For these, fields provide the alternative conceptualisation of geographical phenomena, since they lend themselves naturally to the modelling of continuously varying, multivariate, and dynamic phenomena (Kemp, 1997).

A field is defined as a phenomenon which is distributed over geographical space and whose properties are functions of spatial coordinates and, in the case of dynamic fields, of time. Some phenomena can be converted to fields by taking the limit of the value of the phenomenon, for example, the count or frequency, divided by the area over which the count is evaluated, as the latter tends to zero. This process, which features continuous conceptualisation of discrete phenomena, results in continuous density surfaces. Since measurement must be made over defined areas, such density fields have implied scales and must be treated differently from true fields, that is, fields such as elevation whose variables are directly observable at points.

Natural and environmental phenomena are often measured on continuous scales. Temperature and precipitation can be measured to as many decimal places as the measuring instrument allows, creating contoured surfaces. However, many phenomena such as drought, soil, or vegetation are more often measured as belonging to certain categories using discrete scales of measurement (Webster and Beckett, 1968). Fields are discretised by dividing the area into patches or zones, and assuming the variable is constant within each individual zone, to form maps of soil types, lithologic units, or ecotopes, even though reality may be more objectively described as continuously varying (Duggin and Robinove, 1990; Dungan, 1998).

General consensus has developed over the years about which models should be used. When the phenomena being studied are interpreted as consisting of real or predominantly static entities, which can be mapped on external features, then the object models are used. Cadastral data are commonly stored as objects in urban databases, since property ownership and boundaries have to be accurately verified and measured. Forestry management requires data and analysis at the level of forest stands, which are viewed as polygonal objects characterised by boundaries and attributes, although much spatial heterogeneity exists within individual stands. On the other hand, such digital representations of spatial objects do not deal with geographical complexity that may consist of interacting parts, or may display variation at many different levels of detail over space and time; such phenomena may be better modelled with continuous fields. In other words, field models are used for GIS applications where a comprehensive view of spatial variability is required. Without a clear understanding of the underlying spatial distributions, misused data models could lead to discrepancies between data and the reality these data are intended to represent and, even worse, to malfunction of GIS built upon misrepresented data.

However, there are many possibilities where fields and objects either co-exist, or perform complementary roles. Terrain, as a largely continuous field phenomenon, is often disrupted with discontinuities such as cliffs and ridges, and frequently contains discrete entities such as lakes and roads. Most applications use objects to define field domains, such as fields of water flows along object-definable rivers, or the boundaries of study areas. Samples of field variables usually take the form of points, lines, or areas, as they are captured at individual points, along transects, or within finite areas. On the other hand, fields may be introduced to represent the properties of collections of objects, perhaps as densities, or to represent continuously varying phenomena that form the context for discrete objects. Much geo-processing involves transformation from fields to objects or vice versa. For instance, specific characteristics of a field, such as valleys, ridges, and other extremes, may be extracted from an elevation model. Proximity or density fields are often modelled from a discrete set of objects (Couclelis, 1999).

Having developed a preliminary yet realistic account of geographical modelling above, this chapter will examine key concepts related to fields and objects, and their characterisation and manipulation, with fields taking precedence over objects. Some remarks will be made on issues of inaccuracy, randomness, and vagueness to preview the methodological elegance of field-based handling of geographical uncertainties.

2.2 FIELDS

2.2.1 Models

The concept of fields is fundamental in science. The use of fields in geographical information science has its source in physics via the notion of force fields, which are used to describe phenomena associated with forces caused by distributions of matter, charge, and the like upon imaginary unit particles within certain spatio-temporal domains. Gravitational and magnetic fields are typical force fields. A field can be modelled mathematically as a function of spatial (and temporal) location. Thus, phenomena such as air pressure, temperature, density, and electro-magnetic potential are described by functions of space and time. Scalar fields have at each location a scalar (single) value, while vector fields have multiple values at each location, and are often used to represent the magnitude and direction of phenomena such as wind, or the direction of maximum terrain slope.

Mathematically, a function maps every element from the domain of independent variables to exactly one element in its range, if it is single-valued. A field is continuous if its domain is continuous, that is, a compact and connected set or a continuum (Huntington, 1955). The notion of a continuous field function should not be confused with that of a continuous and smooth mathematical function, as continuity does not necessarily lead to smoothness. Moreover, the mapping rule of a function need not be directly computable, that is, definable as an analytical expression of the domain values, as is usually the case with mathematical functions. A rule can also be defined, for example, as an explicit association of individual elements of the domain and the range, that is, a table listing the association between the two sets.

For field values measured by whatever means, an important concept is that of support. A sampled value of the field is related to an element of the support. Consider rainfall data sampled at irregularly distributed sites in a city for the summer. The support of the field is defined by rain gauges placed at individual sites on the surface of the city, which are elements of the support. The field values are mean values averaged from gauge measurements at individual sites.

In more rigorous terms, a field on a given support B is a function $Z(.)$, which assigns every element x_s of the support a corresponding field value $Z(x_s)$ from the value domain. When the value domain is real and one-dimensional, $Z(.)$ is a scalar field. If the support B is continuous, that is, a dense and compact subset of a metric space, there is a metric $\|.\|$ defined on B to express distance: $dis(x_1, x_2)=\|x1-x2\|$, x_1, $x_2 \in$ B.

The measurement of the field consists of a set of numbers denoted usually as lower-case letters $z(x_s)$ $(s=1,\dots,n)$ as opposed to upper-case Zs representing theoretical field values. These measured values $z(x_s)$ can be linked with the field $Z(x)$ by:

$$z(x_s) = \int_s \varphi(x)Z(x)dx \Big/ volume(x_s) \qquad (2.1)$$

where $\varphi(x)$ is a selection function, s denotes the locality around x_s, and $volume(x_s)$ is a quantity associated with the locality of x_s and used for normalisation:

$$volume(x_s) = \int_s \varphi(x)dx \qquad (2.2)$$

As each selection function is related to x_s, a subset of the support, it makes sense to say that $Z(x_s)$ is the value of $Z(x)$ at x_s. In many applications, the sampled field value $z(x_s)$ is a weighted average of the field $Z(x)$ around the element x_s, where weights are given by a proper selection function. If x_s is a point and the measurement retrieves the field value exactly at that point, then the selection function is a Dirac distribution centred at x_s:

$$\varphi(x) = \delta(x_s - x) \qquad (2.3)$$

In the case of remotely sensed images, the element x_s is a square pixel (although in practice the selection function involves considerable blurring of pixel edges). A simple example is the mean value over x_s with a selection function given by:

$$\varphi(x) = 1/volume(x_s), x \in x_s \text{ and zero otherwise} \qquad (2.4)$$

where $volume(x_s)$ is the integral of dx over the locality of x_s, as shown in Equation 2.2.

Depending on what schemes are adopted for sampling, fields may be modelled with irregular points, regular points, contours, polygons, grids, or TINs (triangulated irregular networks), as illustrated in Figure 2.1. Two types of variables are involved in field models: categorical and continuous. Categorical variables are measured on discrete scales (nominal and ordinal), while continuous variables are measured on continuous scales with interval or ratio properties (Stevens, 1946). Categorical variables are exemplified by land cover and soil type, while examples of continuous variables include atmospheric pressure, bio-mass, crop yields, elevation, slope gradient, rainfall, and air temperature.

There are four field models for categorical variables of which two are used for areas: the grid model, where variation is described by determining the variable's value within each rectangular cell; and the polygon model, where the plane is divided into irregular polygons. For both models, spatial variation within cells and polygons is ignored. Besides, a polygon often removes some of the geometric complexity of boundary lines. This is evident in the classification of remotely sensed images, where each pixel is assigned a single dominant class, and where contiguous patches of pixels with identical classes are further smoothed to generate polygonal maps and to remove unsightly 'jaggies' that result from following pixel edges. Therefore, uncertainties occur in using the grid or polygon models for categorical variables when spatial variations within cells and polygons are generalised, allowing only for a single and dominant category for each cell or polygon.

Field models	Graphical representations	Valid types of variables	Examples
irregular points	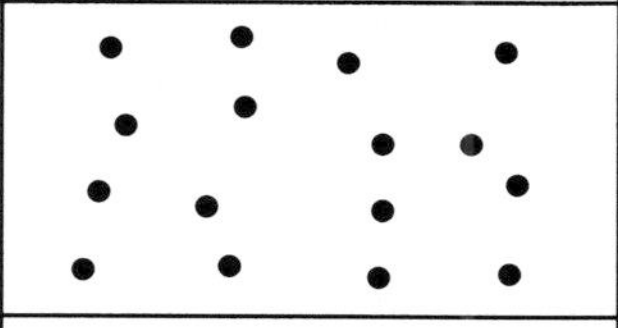	continuous categorical	weather station data
regular points	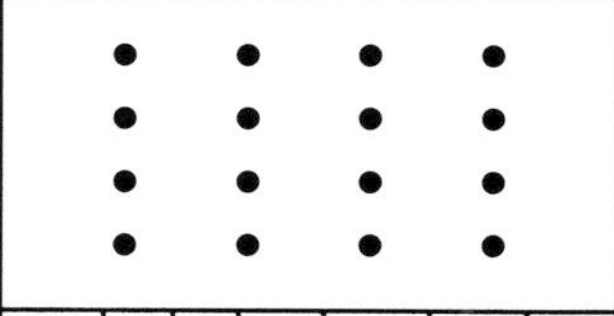	continuous categorical	laser profiling data
contours	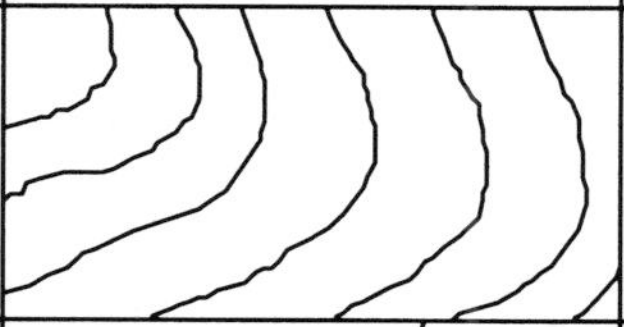	continuous	noise level data
polygons		continuous categorical	land cover data
grids	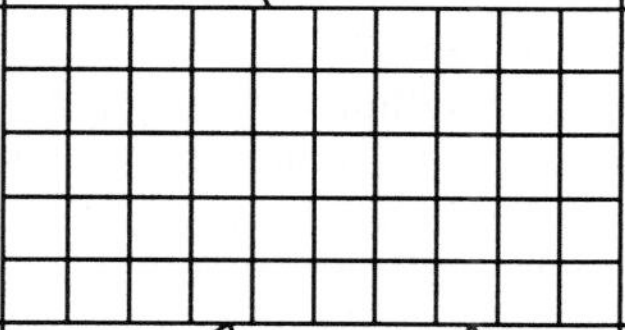	continuous categorical	remotely sensed data
triangulated irregular networks (TINs)	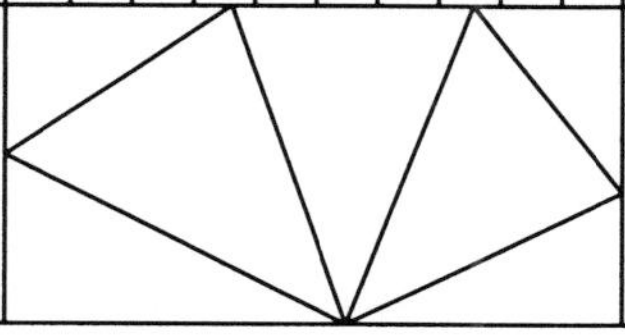	continuous	topographic data

Figure 2.1 The six field models commonly encountered in GIS.

For continuous variables, all the six models are commonly used. Using a TIN model, only critical points defining local discontinuities are required to model the spatial variation under study (Peucker and Douglas, 1975). Thus, TINs are often use for elevation data, where height discontinuities and breaks of slope along triangle edges may fit well with actual Earth topography.

As shown in Figure 2.1, with polygon and grid models it is possible to query the value of a variable at any point in the plane, whereas, with others, values can be obtained at arbitrarily located points only through the use of an additional process called spatial interpolation. Because interpolation is not standard, and estimating variables such as elevation from a set of (ir)regularly spaced points depends on the interpolation method used, it is more difficult to address the problem of uncertainties for those field-based models that require an interpolation process. This is more so for interpolations based on irregularly spaced points and contours, in which the spatial pattern of uncertainties is highly non-uniform, with lowest levels of uncertainty close to the points and contour lines (Goodchild, 1993). Thus, only grid and polygon models are fully field-based models in this sense.

With fields available in discretised form, it is possible to process them with certain operators or algorithms to derive descriptive statistics or other variables required for the solution of spatial problems. This may involve arithmetic or logical operations on fields individually or, more often, in combination. Such examples of field processing are common in practice. Rainfall data are summarised through the routine calculation of seasonal averages and their spatial variation. Elevation data may be used for the derivation of slopes and aspects, which are increasingly generated from GIS as important components in various applications (Hunter and Goodchild, 1997).

A common numeric approximation *slope(i,j)* to the tangent of slope for grid location $x = (i,j)$ is estimated via:

$$\tan(slope(i,j)) = \left[\left(\partial Z/\partial X\right)^2 + \left(\partial Z/\partial Y\right)^2\right]^{1/2} \tag{2.5}$$

where partial derivatives are estimated from:

$$\partial Z/\partial X = (Z(i,j+1) - Z(i,j-1))/2$$
$$\partial Z/\partial Y = (Z(i-1,j) - Z(i+1,j))/2 \tag{2.6}$$

and where the central grid cell (i,j) is shown in relation to its eight immediate neighbours in Figure 2.2. Other forms of the estimating equations include the elevations of all eight neighbours, and give greater weight to the four immediate neighbours than to the four diagonal neighbours (Longley *et al.*, 2001).

$Z(i+1,j-1)$	$Z(i+1,j)$	$Z(i+1,j+1)$
$Z(i,j-1)$	$Z(i,j)$	$Z(i,j+1)$
$Z(i-1,j-1)$	$Z(i-1,j)$	$Z(i-1,j+1)$

Figure 2.2 A 3 x 3 gridded neighbourhood for variable *Z* centred at coordinates (*i,j*).

Remote sensors readily record multispectral fields of emitted or reflected electro-magnetic radiance from the Earth's surface in raster format. An important output from remote sensing is the classification of land, which actually amounts to evaluating categorical variables over space within some prescribed framework consisting of discrete categorical labels. Land cover refers to the material cover on the surface of the Earth, while land use refers to the cultural use of land. A good example is an area that is forested in terms of land cover, but may be used as habitat for wildlife or as recreational land in terms of land use (Jensen, 1979; Townshend and Justice, 1981).

Classification of land cover and other geographical properties is increasingly computerised with digital sensor data, which are subjected to numeric processing in a statistical classifier. Logical operators such as AND and OR may also be useful, especially when multi-source data are incorporated. Let $U(x)$ stand for a categorical variable evaluated with discrete labels of categories for a pixel x, such as grassland and woodland. Then, given spectral radiance data of b bands as vector fields $Z(x)=(Z_1(x),Z_2(x),...,Z_b(x))^T$, a classifier assigns a discrete category to each individual x by a mapping rule CL such that:

$$U(x) = CL(Z_1(x), Z_2(x),..., Z_b(x))$$ (2.7)

where the classification rule may make use of overlaid multivariate data. Preferably, the rule will also make use of spatial dependence within a neighbourhood centred at pixel $x=(i,j)$, as illustrated in Figure 2.3, in order to take advantage of the information present in texture.

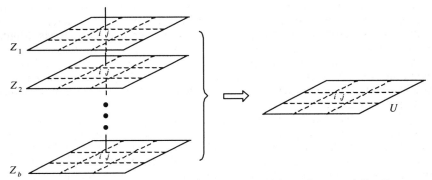

Figure 2.3 Mapping categorical variable U from multiple continuous variables Z.

Remote sensing may also be used to provide cost-effective data on continuous quantities such as biomass, and indices of vegetation strength or vigour. A common vegetation index is the so-called normalised difference vegetation index or NDVI in short. It is based on the difference of reflectance in the near-infrared and red spectral bands, normalised by their sum (Campbell, 1996). Thus, $NDVI(x)$ for a location, or equivalently, pixel x is calculated as:

$$NDVI(x) = (Z_{NIR}(x) - Z_R(x))/(Z_{NIR}(x) + Z_R(x))$$ (2.8)

where $Z_{NIR}(x)$ and $Z_R(x)$ with upper-cased subscripts represent reflectance in the near-infrared and red bands at location x.

In multi-criterion spatial evaluation, different variables are often overlaid to drive hazard prediction or suitability analysis. It seems straightforward to use the notion of multivariate $Z(x)$ in Equation 2.7, and to include layers of co-registered variables such as elevation, slope, land cover, and others to produce information useful to the solution of the problem (Franklin and Woodcock, 1997; Maselli *et al.*, 1998). Many examples are available for consultation in connection with GIS environmental modelling, and this topic is not discussed further here.

No matter what models and methods are used to represent and analyse field variables, uncertainties are bound to exist. Firstly, this is because of the general complexity of geographical fields, implying inaccuracy, or rather the inability of mathematical functions or certain mapping rules to assign unique and correct field values to individual locations. For example, gravitational fields are known to be spatially varying depending on geographical location and altitude, although the Newtonian law has established mathematically precise methods for the calculation of gravity between idealised mass points. Measured gravity values are known to deviate from theoretical ones. But the fact is that, to geodetic and geophysical professionals, any formulation by models is smart approximation at best.

Further generalisation and simplification come about in the context of GIS-supported spatial problem-solving, where spatial variations of field variables are necessarily discretised, mostly into finite grid cells, to facilitate efficient computer processing. With such pre-processing, continuous variables are typically smoothed versions of more rugged reality, while individual categories are homogenised, with inter-categorical heterogeneity often over-sharpened to depict categorical variables as choropleth or colour-mosaic maps.

Whether the variable is continuous or discrete, geographical fields are measured by a finite number of samples or observations. The density of sampling determines the spatial resolution of the mapped field, and much research has been conducted on optimum sampling, such as for the creation of DEMs (digital elevation models) (Balce, 1987). For instance, samples taken every hour will miss shorter-term variation, while samples taken every 1km will miss any variation at resolutions less than 1km. For irregular spatial sampling, point samples miss variation between points, and samples taken along transects or contours miss variation between lines.

Clearly, characterisation and documentation of source data in terms of resolution and spatial aggregation is important before embarking on any geo-processing, if sensible use is to be made of field variables. There would be no gain of accuracy at all by spuriously refining spatial resolution or pixel size for a gridded field previously sampled with coarser resolution. Contours provide rather unbalanced representation for a field, as an infinity of points along individual contours reveals limited information for locations between them. On the other hand, increasing spatial resolution by sampling fields at finer intervals tends to provide more information regarding the underlying spatial variation, and perhaps more accuracy. But, there is often some sort of compromise to be made between resolution and accuracy, in particular for global databases (Tobler, 1988).

When dense and repetitive sampling is possible or an affordable luxury, one may become disillusioned with the additional variation revealed at finer resolution. But such spatial variability should be taken as one of the intrinsic properties of geographical fields, rather than any indication of their unpredictability. A further characteristic of fields is spatial dependence, since the values of the field at individual locations tend to be correlated. By Tobler's 'law of geography', things closer to one another tend to be more similar than those further apart (Tobler, 1970).

Spatial dependence is fundamental to many geographical distributions and their interpretation. For field models such as contours and irregular points that do not fully sample an underlying variable, querying at unspecified or unsampled locations requires a process of spatial interpolation. In this process, implicit use is made of spatial dependence so that complete field models can be constructed with reasonable accuracy. Even for fields completely specified by grid or polygon models, spatial dependence is equally important for reliability or meaningfulness in derivative variables created from them, since derivative data may be flawed if they do not accommodate spatial dependence (although many computer algorithms perform geo-processing routinely without accounting for spatial dependence). Lastly, if the spatial variability and dependence of a field variable are known, optimal choices can be made about sampling, representation, and analysis in terms of spatial resolution. Samples can be configured to adapt to the inherent spatial variability and dependence of the underlying variables, by placing samples more densely in highly variable areas than in areas with relative homogeneity.

In short, fields are potentially powerful for capturing continuously varying properties of many spatially distributed phenomena. For successful field-based geographical modelling, the spatial variability and dependence inherent in geographical fields have to be modelled. The next section establishes the core role played by geostatistics in the characterisation and quantification of spatial dependence, which is crucial for the sensible handling and understanding of geographical information.

2.2.2 Spatial Variability and Dependence

As was mentioned in the preceding section, it is important to recognise that many real-world phenomena exhibit random variation, due not only to inherent variation in the phenomenon, but also to measurement inaccuracy and other forms of uncertainty. To account for such uncertainties, probabilistic approaches model each field value as a random variable. Every $x \in D$, where D is the domain of the field, is mapped to a random variable $Z(x)$. The cumulative probability distribution function $F(z,x)$ is used to characterise the field, and is defined by (again for a scalar field):

$$F(z,x) = P(Z(x) \leq z) \tag{2.9}$$

where z indicates a specific field value.

A random field needs a set of numbers to represent the field value at location x. Descriptive measures such as mean values and standard deviations are often used. Sampling of the phenomenon may provide enough information to describe $Z(x)$. An example is the characterisation of $Z(x)$ using its empirical distribution function (histogram) with c classes:

$$p_k = P(z_k < Z(x) \leq z_{k+1}) = F(z_{k+1}, x) - F(z_k, x), k = 1,...,c, \ z_1 = -\infty, \ z_{c+1} = \infty \tag{2.10}$$

Such a probabilistic formalism provides the basis of geostatistical analysis and the geostatistical exploration of geographical data (Journel, 1986; Cressie, 1993; Journel, 1996). In geostatistics, geographies of interest are modelled as spatially correlated random variables. As described above, a random variable is a variable that can take a variety of outcome values according to some probability (frequency) distribution. Apart from being location-dependent, a random variable is also information-dependent, because its probability distribution tends to change as more data become available. A random variable is usually denoted by Z, while z often indicates a measured or observed value of Z. Examples include pollutant concentration over a marshland and population density in a municipal region.

The basic model in geostatistics is:

$$Z(x) = m_Z(x) + \delta(x) + \varepsilon \tag{2.11}$$

where $Z(x)$ is the value of variable Z at point x, $m_Z(x)$ is a deterministic function describing the general structural component of variation and may be taken as the mean field of Z, $\delta(x)$ describes the spatially correlated local random variation, and ε is the random (or noise) term.

Under the assumption of the so-called *intrinsic* hypothesis, there is a stationary variance of the differences in a random variable between locations separated by a given distance and direction. The semi-variance of difference (usually denoted by γ) is half of the expected (E) squared difference between two values:

$$\gamma_Z(h) = E[(Z(x_1) - Z(x_2))^2]/2 \text{ for } x_2 - x_1 = h \tag{2.12}$$

where $Z(x_1)$ and $Z(x_2)$ are the values of variable Z at locations x_1 and x_2 respectively, which are separated by lag h. For spatially isotropic variation, the lag h is simplified as distance and, for this reason, the condition specified in Equation 2.12 can be relaxed as $\|x_2 - x_1\| = \|h\|$.

The function that relates γ to h is called the semivariogram, and is the function used to quantify the spatial correlation or dependence. An experimental semivariogram $\gamma_Z(h)^*$ can be estimated from sampled data via:

$$\gamma_Z(h)^* = \frac{1}{2N(h)} \sum_{i=1}^{N(h)} (z(x_1) - z(x_2))^2 \text{ for } x_2 - x_1 \approx h \tag{2.13}$$

where $N(h)$ is the number of pairs of observation points approximately separated by lag h. By changing h, a set of values is obtained, from which the experimental semivariogram is constructed.

The experimental semivariogram depends on the scale of the survey and the support of the sample, and its accuracy relies on the sample size. In developing semivariograms, it is necessary to make some assumptions about the nature of the underlying variables. When observed variations do not reveal significant anisotropy, that is, directional variation in spatial dependence, the sample points that are usually irregularly distributed can be arranged as pairs separated by a range of distances, between distance 0 and the maximum distance in the study area. For every pair of points, one computes distance and the squared difference in z values. Then, one assigns each pair to one of the distance ranges, and accumulates total variance in each range, which is averaged and plotted at the midpoint distance of each range to recover an experimental semivariogram.

As is commonly understood, experimental semivariograms may be subject to various errors. However, in principle, the true or theoretical semivariogram should be continuous. Thus, experimental semivariograms are usually fitted with suitable mathematical models, which may be exponential, Gaussian, spherical, or power (Cressie, 1993). Figure 2.4 shows, as an example, an experimental semivariogram (the scatter plot) fitted mathematically with a model (the solid line).

There are several features to note in the semivariogram. At relatively short lag distances h, the semi-variance is small but increases as the distance between the pairs of points increases. At a distance referred to as the range, the semi-variance reaches a relatively constant value referred to as the sill, as illustrated in Figure 2.4. This implies that beyond this distance, the variation in z values is no longer spatially correlated (in practice it is often difficult to avoid subjective judgment in the determination of the range, as in the example, where the curve begins to flatten out at $h=30$ but is still increasing at $h=60$). The experimental semivariogram often cuts the ordinate at some positive value, termed the nugget effect or variance, as shown in Figure 2.4. This might arise from measurement errors, spatially dependent variation over distances much shorter than the smallest sampling interval, or from spatially uncorrelated variation.

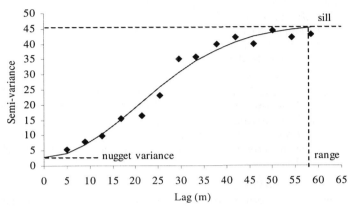

Figure 2.4 An illustrative experimental semivariogram (diamonds) and its model (solid line).

One can check the existence and magnitude of measurement errors by carrying out repeated measurements or sampling of the same variable at the same location. Usually this is done for several locations to reduce possible bias. Measurement errors affect the estimation of semivariograms. However, as measurement errors are assumed to be non-correlated and should be the same for all data points, they should not affect the overall form unless it is large relative to the total variance, and unless the number of sample points is very limited (Burrough, 1995, pers. comm.).

Another quantity of spatial variability is spatial covariance, $\text{cov}_Z(h)$, which is defined as the expectation of the product of a variable's deviates from local means for locations separated by a lag h:

$$\text{cov}_Z(h) = E[(Z(x_1) - m(x_1))(Z(x_2) - m(x_2))] \text{ for } x_2 - x_1 = h \tag{2.14}$$

As for semivariograms, covariance can be estimated experimentally with sample data by:

$$\text{cov}_Z(h)^* = \frac{1}{2N(h)} \sum_{i=1}^{N(h)} (z(x_1)z(x_2) - m^2) \text{ for } x_2 - x_1 = h \tag{2.15}$$

where $N(h)$, $z(x_1)$, and $z(x_2)$ follow similar definitions to those in Equation 2.13.

It is further possible to derive from $\text{cov}(h)$ a correlation parameter ρ as:

$$\rho_Z(h) = \text{cov}_Z(h)/\text{cov}_Z(0) \tag{2.16}$$

where $\text{cov}_Z(h)$ is already defined in Equation 2.14, and $\text{cov}_Z(0)$ refers to the variance of the variable Z and serves to standardise spatial covariance at lag h.

As a stationary property, $\gamma_Z(h)$ is linked with the difference of covariances by:

$$\gamma_Z(h) = \text{cov}_Z(0) - \text{cov}_Z(h) \tag{2.17}$$

and it is straightforward to express Equation (2.16) another way:

$$\rho_Z(h) = 1 - \gamma_Z(h)/\text{cov}_Z(0) \tag{2.18}$$

Semivariograms include both lag-to-lag and local trend variabilities, while spatial covariances tend to filter out the effects of mean and variance trends. For this reason, semivariograms and spatial covariances should be used in combination to provide interpretation of spatial dependence structures, as discussed by Rossi *et al.* (1994). Specific discussion of spatial dependence in remotely sensed images is provided by Curran (1988) and Jupp *et al.* (1988).

As an extension to Equation 2.14, the cross-covariance $\text{cov}_{Z1Z2}(h)$ for two variables $Z1$ and $Z2$ may be defined as:

$$\text{cov}_{Z1Z2}(h) = E[(Z1(x_1) - m_{Z1})(Z2(x_2) - m_{Z2})] \text{ for } x_2 - x_1 = h \tag{2.19}$$

where m_{Z1} and m_{Z2} stand for the means of variables $Z1$ and $Z2$, respectively. The cross-covariance function $\text{cov}_{Z1Z2}(h)$ quantifies the similarity between the two variables $Z1$ and $Z2$ at varying distances apart.

Similar to Equation 2.16, the cross-covariance $\text{cov}_{Z1Z2}(h)$ may also be standardised using the product of the standard deviations of $Z1$ and $Z2$, that is, $\text{cov}_{Z1}(0) \, \text{cov}_{Z2}(0)$, by:

$$\rho_{Z1Z2}(h) = \text{cov}_{Z1Z2}(h)/(\text{cov}_{Z1}(0)\text{cov}_{Z2}(0))^{1/2} \tag{2.20}$$

The standardised form of the spatial correlation quantity ρ is also usefully related to spatial autoregressive models, as discussed by Cliff and Ord (1973), and later by Goodchild (1980), Haining *et al.* (1983), Anselin (1993), and Cressie (1993). It has to be noted that the parameter ρ, which will be defined in the discussion of spatial autoregression below, measures the overall strength of spatial dependence in a field variable, rather than lag-dependent strengths as in Equation 2.16.

Suppose a set of field values exists in the form $Z = (Z(x_1), Z(x_2), ..., Z(x_n))^T$ for n locations specified on the plane (^{T} stands for matrix transpose). Let ε be a joint n-dimensional Gaussian distribution with mean 0 and diagonal variance matrix $\text{cov}_{nxn} = \sigma^2 I$, where σ^2 is variance and I stands for an identity matrix of order n. Let W denote a matrix indicating spatial dependence such that w_{ij} ($i, j = 1, 2, ..., n$) is set to be greater than 0 if $Z(x_i)$ is thought to be positively correlated with $Z(x_j)$, otherwise w_{ij} is set to 0. Also, w_{ii} must be set to 0 for the inverse of matrix $(I - \rho W)$ to exist. A spatial regression model is defined as:

$$(I - \rho W)(Z - E[Z]) = \varepsilon \tag{2.21}$$

or equivalently as:

$$Z = E[Z] + \rho W(Z - E[Z]) + \varepsilon \tag{2.22}$$

where $E[Z]$ stands for the expectation of Z.

Clearly, Equation 2.22 is reminiscent of the Markovian model for spatial dependence (Besag, 1974). For a brief introduction to Markovian models for spatial dependence, a neighbourhood N_s is defined as consisting of all locations x_i neighbouring a specific location x_s, in which $Z(x_s)$ depends functionally on $Z(x_i)$:

$$N_s = \{x_i, x_i \text{ is a neighbour of } x_s \text{ but } i \neq s\} \text{ for } i, s = 1, 2, ..., n \tag{2.23}$$

Based on this, it is easy to see that the slope defined in Equation 2.5 implies a neighbourhood of four grid cells sharing sides with the central cell at (i, j). A field is said to be a Markov random field if its conditional probabilistic distribution defines a neighbourhood structure such as N_s in Equation 2.23. Further discussion about geostatistical and Markov models for describing spatial dependence and geographical fields will be provided in Chapter 5, but Griffith and Layne (1999) may be consulted for insightful understanding and real data analyses across the breadth of spatial statistical data handling.

The foregoing exposition has been focused implicitly on field variables of continuous properties. Categorical variables such as soil and vegetation constitute the other type of geographical fields, which are typically depicted as contiguous groups of grid cells or polygons that are distinctively coloured. Consider a set of field values Z measured on a binary scale such that element $Z(x_s)=1$ if a category occurs at location x_s, 0 otherwise ($s=1,2,...,n$). A useful statistic for spatial dependence in such categorical occurrences was proposed by Moran (1948), and is also known as Moran's contiguity ratio, defined as:

$$I_{Moran} = Z^T W Z / Z^T Z \tag{2.24}$$

where matrix W follows the same definition as in Equation 2.21.

2.2.3 Spatial Interpolation

The spatial dependence that is characteristic of many geographical fields allows one to make inferences about a variable and its spatial variation in the light of measured data. Spatial interpolation is the procedure of estimating the value of a field variable at unsampled sites within the area covered by sampled locations. In other words, given a number of samples whose locations and values are known, such as weather station readings, spot heights, or oil well readings, spatial interpolation is used to determine the values of the underlying variable at other locations.

For mapping categorical variables, proximal methods are used, by which all values are assumed to be equal to the nearest location of known categories, and the output data structure is Thiessen polygons that partition space into influence zones centred at known locations. Thiessen polygons have ecological applications and are naturally interpretable for categorical data, although they were used originally for computing areal estimates from rainfall data measured on continuous scales (Thiessen, 1911).

Interpolators may be differentiated by various criteria. Global interpolators determine a single function for a whole region, while local interpolators apply an algorithm repeatedly to portions of the study area. A change in an input value only affects local areas for a local interpolator, but a change in one input value affects the entire map if it has been globally interpolated. Exact interpolators honour the data points, such that the surface generated passes through all points whose values are known. On the other hand, approximate interpolators are used when there is some uncertainty about the given surface values. Another important distinction among interpolators is between those that are deterministic and those that are stochastic. Stochastic methods incorporate the concept of randomness, and the interpolated surface is conceptualised as one of many that might have been observed, all of which could have produced the known data points.

As a kind of deterministic interpolator, moving-average or distance-weighted-average techniques estimate the field value $z(x)$ at an unknown location x as a distance-weighted average of the values at the n sampled locations:

$$z(x) = \sum_{s=1}^{n} wt(x, x_s) z(x_s) \bigg/ \sum_{s=1}^{n} wt(x, x_s) \qquad (2.25)$$

where wt is some function of distance, such as $wt(x,x_s)=1/\|x-x_s\|^k$, with k often set at 1 or 2. Many variants are possible with moving average interpolators by using different distance functions, or by varying the number of points used.

Trend surface analysis performs a kind of stochastic interpolation, in which the unknown $z(x)$ at a location $x(X,Y)$ is approximated by a polynomial using powers of X and Y. Thus, a linear equation (degree 1) describes a tilted plane surface, while a quadratic equation (degree 2) describes a simple hill or valley. Given sampled data $z=(z_1,z_2, \ldots, z_n)^{\mathrm{T}}$, a trend surface:

$$z(x) = F_{XY}(x)\beta + \varepsilon \qquad (2.26)$$

is to be solved for coefficients β (column vector), where $F_{XY}(x)$ stands for a row vector consisting of powers of X and Y, and ε is typically a noise term of zero mean and constant variance. The polynomial coefficients are estimated by least squares as:

$$\beta = \left(F_{XY}^{\mathrm{T}} F_{XY}\right)^{-1} F_{XY}^{\mathrm{T}} z \qquad (2.27)$$

Trend surface analysis amounts to a global interpolator, with a relatively light computing load. But trend surface analysis assumes that the general trend of the surface is independent of random errors found at each sampled point. This is rarely the case in practice, and it is consequently very difficult to separate a general trend from local, non-stationary effects. A further fundamental restriction for trend surface analysis lies in the assumption of spatially uncorrelated errors.

Kriging, which was developed by D.G. Krige (1962) as an optimal method of interpolation for use in the mining industry, and by G. Matheron (1963) as the theory of regionalised variables, has been widely accepted as a method of spatial interpolation by spatial information specialists. Growing from the need to interpolate certain properties (or attributes) of interest from sparse data, Kriging outperforms other spatial interpolators by modelling properties as spatially correlated random variables within a rigorous theoretical framework (Delfiner and Delhomme, 1975).

In Kriging, a constant local mean and a stationary variance of the differences between places separated by a given distance and direction are usually assumed. The stationarity in spatial distributions is expressed by semivariograms showing how the average difference between values at pairs of points changes with the distance between the points. Semivariograms are used further for weighting individual sample points in the neighbourhood of the locations to be estimated.

Suppose as before that data $z(x_s)$ exist at sampled locations x_s ($s=1,2,\ldots,n$). Kriging predicts the field value at another location x where the value $z(x)$ is not known using a weighted sum of values. This is expressed as:

$$z(x)^* = \sum_{s=1}^{n} \lambda_s z(x_s) \qquad (2.28)$$

where λ_s stands for the weight attached to the sample located at x_s.

With assumptions of stationary field mean and spatial covariance, weights λ are selected so that the estimates are unbiased (if used repeatedly, Kriging would give the correct result on average) and have minimum variance (the variation between repeated estimates at the same point is minimum). These two requirements lead to:

$$\sum_{s=1}^{n} \lambda_s = 1 \tag{2.29}$$

and minimisation of the variance of the estimation error $z(x)-z(x)^*$, which is expressed as:

$$E[z(x) - z(x)^*]^2 = E[z(x) - m_Z]^2 + \sum_{s=1}^{n}\sum_{q=1}^{n} \lambda_s \lambda_q \mathrm{cov}_Z(x_s - x_q) - 2\sum_{s=1}^{n} \lambda_s \mathrm{cov}_Z(x - x_s) \tag{2.30}$$

where $E[z(x) - m_Z]^2$ is the total variance of the variable Z, with both s and q indicating sampled locations from the set $\{x_1, x_2, \ldots, x_n\}$.

Minimising the variance of the estimation error expressed in Equation 2.26 under the condition set forth in Equation 2.29 leads to Kriging weights and hence to the Kriged estimate $z(x)^*$. The Kriging variance, $\sigma_{Z(x)}^2$, is given by:

$$\sigma_{Z(x)}^2 = 2\sum_{s=1}^{n} \lambda_s \gamma_Z(x - x_s) - \sum_{s=1}^{n}\sum_{q=1}^{n} \lambda_s \lambda_q \gamma_Z(x_s - x_q) \tag{2.31}$$

It is clear that Kriging is a procedure that is much more complicated than moving average or trend surface analysis. However, Kriging is based on the exploitation of explicitly measured spatial dependence for weighting sample data within an unknown location's neighbourhood, and Kriging not only generates estimated field values at unsampled locations but also provides information about their error variances. These geostatistical properties are well worth exploring, especially when knowledge of uncertainty is sought.

2.3 OBJECTS

2.3.1 Models

The preceding section discussed field variables, for which spatial variability and dependence are seen as important characteristics that can be effectively modelled using geostatistics, and quantified in terms of spatial auto- and cross-correlation. On the other hand, much spatial problem-solving concerns individual entities and their spatial interaction. In such cases geographical phenomena are more readily conceptualised as discrete objects, with geo-relational models for data representation and analysis. For example, transportation planning and management involve key locations or places, intersections, roads, streets, rivers, canals, and bridges, which are themselves real-world entities and thus naturally modelled as points, lines, or areas. These objects are indexed by positional descriptions and often associated with qualitative and quantitative facts such as node annotations, numbers of street lanes, bridge spans, and speed limits. Not surprisingly, map data are routinely digitised as discrete points, lines, and areas. While many modelled objects are real entities, others are abstracted artefacts. Notable examples include meteorologically delineated weather fronts represented as lines, and atmospheric depressions represented as points.

Under an object-based model, spatial data exist in the form of discrete points, lines, and areas. Points occupy very small areas in relation to the scale of the database (e.g., towns and cities on small-scale maps). Lines are used for real linear phenomena, and are exemplified by entities such as highways, rivers, pipelines, and power lines. Area objects are used to represent distributions such as building blocks, census tracts, lakes, land parcels, plantations, and other real or abstracted spatial entities that occupy areas at the scale of a GIS.

For a formal introduction to object-based models, a few concepts such as entities, objects, features, and attributes are defined here. An entity is a real-world phenomenon that is not subdivided into phenomena of the same kind. An object is a digital representation of all or part of an entity. A feature is a defined entity and its object representation. Further, an attribute is defined as a characteristic of an entity, and an attribute value is a specific quality or quantity assigned to an object. Of course, objects must be associated with position, which is usually specified by a set of coordinates, postal codes and addresses, and other specialised addressing systems such as those used for highway geo-referencing (Noronha *et al.*, 1999). It becomes clear that an object-based data model expresses the real world as consisting of objects associated with their attributes. The National Committee for Digital Cartographic Data Standards (NCDCDS) (1988) published a classification of cartographic objects, which were usefully extended into a framework for object-based geographical modelling by Clarke (1995). An illustration with adaptation is presented as Figure 2.5.

First, as shown in Figure 2.5, points and nodes are the two types of generic point objects, where points include entity points, label points, and area points. An entity point is used principally for identifying the position of a point entity such as a tower or a historical landmark. A label point is used principally for displaying map and chart text to assist in feature identification. An area point is a point within an area carrying attribute information about that area. A node is a topological junction or end point that may specify geometric position.

Secondly, for lines, there are line segments, strings, arcs, links, chains, and rings. A line segment is a direct line between two points. A string is a sequence of line segments that does not have nodes, node IDs, left/right IDs, nor any intersections with itself or with other strings. An arc is a locus of points, which form a curve defined by a mathematical function such as a spline. A link is a connection between two nodes, while a directed link is a link with one direction specified. A chain is a directed sequence of non-intersecting line segments or arcs with nodes at each end. It may be a complete chain having node IDs and left/right area IDs, an area chain having left/right area IDs but no nodes IDs, or a network chain having nodes IDs but no left/right area IDs. A ring is a sequence of non-intersecting chains, strings, links, or arcs with closure. It represents a closed boundary, but not the interior. As an example, a ring created from links is shown in Figure 2.5.

Lastly, an area object is a bounded continuous two-dimensional object that may or may not include its boundary. Area objects may be irregularly or regularly shaped (usually rectangular). Area objects of irregular shape include interior areas and polygons. An interior area does not include its boundary. A polygon consists of an interior area, one outer ring, and zero or more non-intersecting, non-nested inner rings; a simple polygon has no inner rings, while a complex polygon has one or more inner rings. There are two types of rectangular area objects: pixels and grid cells. A pixel is a two-dimensional picture element that is the smallest non-divisible element of an image. A grid cell is a two-dimensional object that represents an element of a regular or nearly regular tessellation of a surface.

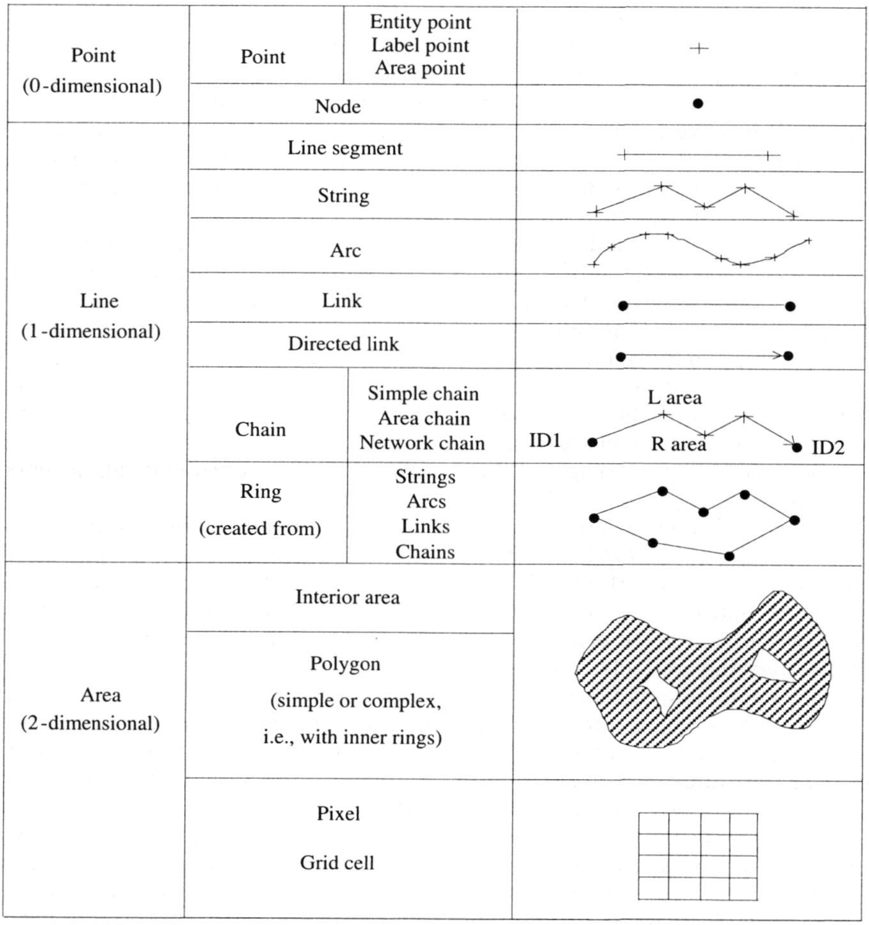

Point (0-dimensional)	Point	Entity point / Label point / Area point	+
	Node		●
Line (1-dimensional)	Line segment		
	String		
	Arc		
	Link		
	Directed link		
	Chain	Simple chain / Area chain / Network chain	
	Ring (created from)	Strings / Arcs / Links / Chains	
Area (2-dimensional)	Interior area		
	Polygon (simple or complex, i.e., with inner rings)		
	Pixel / Grid cell		

Figure 2.5 NCDCDS classification of cartographic objects and their diagrammatic illustration.

   In this conventional model of exact objects, spatial entities are represented with crisply delineated points, lines, areas, and volumes in a defined and absolute reference system. Lines link a series of exactly known points, areas are bounded by exactly defined lines, and volumes are bounded by smooth surfaces. Networks are formed from lines linked into a defined topology. Thus, an open network can represent rivers, while a closed one may define the boundaries of polygons standing for such phenomena as land parcels, census tracts, or administrative areas. Cartographic convention has reinforced the object-based formalism for geographical abstraction by mapping objects with points and lines of given shapes and sizes.

   Whatever form an object may take, it must have an identity making it an individual differentiable from any other. Objects are usually defined with sharp boundaries and have attributes that are independent of location or position. Here, a distinction is made between objects and fields, as grids and polygons are used as models for both fields and objects, causing potential confusion.

For field variables, the properties of the space within polygons are described homogeneously, and the values specific to individual polygons are assumed to be constant over each polygon's total extent, as is the case with the choropleth model. In general, however, it is to be expected that variation will occur even within polygons that are depicted as homogeneous. For this, field-based models provide mechanisms that permit evaluation and incorporation of spatial variability within individual polygons or grid cells, perhaps by retaining more source information.

For object-based modelling, on the other hand, polygons and grid cells are used differently from fields. Grid cells are merely artefacts in geo-processing, and may be directly linked to discrete polygons with some inaccuracy in boundaries. Attributes should be considered as properties of each object in its entirety. For example, land parcel boundaries are accurately surveyed and the attributes of ownership, taxable values, and so on apply uniformly to the whole object. It would be difficult to argue about how much tax discrete points lying within a parcel would have to pay to county revenue.

Data modelling in computer science has evolved from hierarchical, network, and relational to object-oriented methods. Geographical things and places are often organised in different hierarchies, and a road atlas or an institutional entity can be easily fitted within hierarchical, tree-like structures where leaves record data at tiered levels. In a network model, relationships among entities are structured with links. For instance, polygons may contain sequential pointers to arcs, which are in turn linked to coordinate points. Relational models use tables to store entities and their attributes, and rely on comprehensive indexing for queries and data manipulation by joining table entries through common identifiers. Relational data models in the extended form of geo-relational data models have secured a sound place in many GIS software systems, including ARC/INFO (Morehouse, 1985). As single models rarely achieve optimal efficiency of storage and processing, it is often hybrid solutions that survive testing and updating.

Concepts such as objects, classes, attributes, and relations have found natural acceptance in GIS, where objects can be constructed from geometric primitives as the basis for geographical representation (Molenaar, 1998). New developments in data modelling in computer science are perhaps most evident in the object-oriented approach to programming and database design (Coad *et al.*, 1995). In this comparatively new approach, a class of objects shares certain attributes, measures, algorithms, and operations that are applicable only if an object meets the class's formal specifications. In the object-oriented approach, a class of objects implies a structure into which are encapsulated certain attribute templates and algorithms, and which are passed on to any particular object or instance in the class. Thus, Highway 101 should possess, by inheritance, all the attributes defined for the class of highways. Class-specific measures of positional errors should also find a home in object-oriented structures, so that map users are informed of varying levels of inaccuracy in digitised roads and coastlines. Further, simple objects are joined or merged to form complex ones, while a large object may be split into smaller ones. An object may be seen as falling inside or outside another. Obviously, because of the separability of position and attributes, objects can be modelled and manipulated in an object-oriented way with more simplicity and flexibility than otherwise.

As has been seen so far in this section, an object-based model is suitable for well-defined entities such as roads, buildings, and land parcels, and is widely used in cartography and facilities management. Spatial information technologies such as land surveying equipped with total stations, or GPS and photogrammetric triangulation, may be used to provide precision measurement of positional data along with unambiguous identification and attributing of certain discrete objects. This process is effectively dominated by the human power of interpretation and delineation, and is widely believed to be of acceptable accuracy for many spatial applications.

However, it is important to note that representing well-defined entities by using object-based models is considered suitable only under certain conditions, because pure points, lines, and areas do not exist in the real world. Besides, entities have varying and irregular sizes and shapes. Thus, it is rare to find an object-based model that duplicates the real world accurately. Moreover, cartographic generalisation often leads to simplified and smoothed maps, either as the technical consequences of procedures, or for aesthetic purposes (Douglas and Peucker, 1973; MacEachren, 1994). The difference between the modelled and real worlds represents the uncertainty due to geographical abstraction. Such uncertainty is complicated by the limitations inherent in the process of spatial data acquisition, by inaccuracies in measuring position and in evaluating attributes for objects. These topics will be discussed in the next chapter.

2.3.2 Some Geometric and Topological Operations

The formalism of objects in terms of position and attributes contributed greatly in the early developments of digital spatial databases, and in the much more ambitious establishment of national spatial data infrastructures oriented to data sharing. Geo-relational data models and Euclidean geometry have played important roles ever since paper maps were digitised into alpha-numeric versions, and methods for map-based spatial measurement and data handling seem to fit well with the transition of map data handling from analogue to digital media (Maling, 1989). Functions such as the measurement of the lengths and areas of geographical features, and the complex editing processes of computerised cartography, provided evidence for the advantages of GIS over more traditional analogue approaches, which are inefficient and may be subject to a host of technical limitations and human errors.

The positional and attribute dimensions of discrete objects are largely independent of each other, as was shown in the previous section. With object-based modelling, spatial entities are distinct from non-spatial things or concepts due primarily to positional descriptions. On the other hand, collection and validation of attribute data are likely to fall within the responsibilities of, or at least to require the co-operation of specialists who undertake or contract for such GIS applications as emergency routing or education planning, because of the need for local knowledge and expertise. Clearly, such attribute-related issues are largely outside of position fixing, provided that objects stored in spatial databases are unambiguously referenced and recoverable as unique entities or places upon field inspection. It is therefore both reasonable and sensible to focus on object positioning. This section will look into some geometric and topological manipulation of points, lines, and areas with emphases on distance and area calculation.

Suppose a series of points are positioned at x_1 through x_7, as shown in Figure 2.6(a). In a two-dimensional space, positional descriptions for point objects are defined by coordinate pairs (X,Y). A line segment is easily defined by its two end points, such as x_1 and x_2 shown in Figure 2.6(b), where a polyline L is built from the series of points.

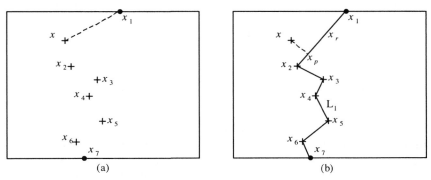

(a)   (b)

Figure 2.6 Distance calculation: (a) between points; and (b) between a point and a line.

Given coordinates for individual points, the calculation of distances between any two points seems straightforward using the Euclidean distance norm. For example, the distance between x and x_1 is written as:

$$\|x - x_1\| = \left[(X - X_1)^2 + (Y - Y_1)^2 \right]^{1/2} \tag{2.32}$$

where (X,Y) and (X_1,Y_1) stand for the coordinate pairs of points x and x_1, respectively. The calculation of line length for L_1 is simply a sum of $\|x_i - x_{i+1}\|$, where subscript i ranges from 1 to 6 for the example shown in Figure 2.6(b).

While distance between specified points is calculated easily, determination of the distance between x and L_1 is less obvious, since distance between a point and a line is defined often as the minimal separation between them. For the latter situation, point x_p, at which the line L_1 has the minimal distance from x, needs to be identified beforehand. Figure 2.6(b) shows that vector (x,x_p) is perpendicular to line segment (x_1,x_2). The minimal distance between x and L_1 is reduced to $\|x - x_p\|$, which is calculated using the notation of vector products as:

$$\|x - x_p\| = \left\| (x - x_1) \times (x_2 - x_1) / \|x_2 - x_1\| \right\| \tag{2.33}$$

A more useful representation for lines is to view them as consisting of an infinite number of points x_r lying on the line segment bounded by two end points, say x_1 and x_2, as shown in Figure 2.6(b). It is possible to write a linear combination of two end points as:

$$x_r = (1 - r)x_1 + rx_2 \tag{2.34}$$

where scalar r represents the ratio of the distance between x_r and x_1 to that between x_2 and x_1. r must be a positively valued real number bounded by 0 and 1, so that x_1 and x_2 can themselves be obtained from Equation 2.34 by taking r values of 0 and

1, respectively.

    With clear specification of the end points of individual line segments, it is possible to use the line representational method in Equation 2.34 to traverse all continua of points along the length of polyline L_1. Extension to polygon boundaries is not difficult. Thus, the distance between points, lines, and areas may be formulated by solving for the minimisation of $\|x - x_r\|$, although this is not explored here.

    In addition to distance, areas are often evaluated for their areal extents. Consider the polygon A shown in Figure 2.7, whose boundaries are defined by a closed sequence of line segments x_1 to x_2, x_2 to x_3,...,x_n to x_1. For simplicity, x_1 is also identified as x_{n+1}, while point x_n is taken as x_0.

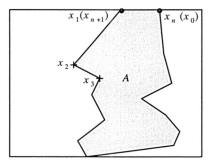

Figure 2.7 Calculation of the areal extent for polygon A.

    A well-known formula is developed on vector products. The sum of vector products of points along A's boundary produces the area for A, that is:

$$area(A) = 1/2 \sum_{i=1}^{n} x_i \times x_{i+1} = 1/2 \sum_{i=1}^{n} (X_i Y_{i+1} - X_{i+1} Y_i) = 1/2 \sum_{i=1}^{n} X_i (Y_{i+1} - Y_{i-1}) \qquad (2.35)$$

where (X_i, Y_i) stands for the coordinate pair of point x_i. Points are numbered counter-clockwise to ensure positiveness for area A, as shown in Figure 2.7.

    In map overlay, points, lines, and areas are overlaid to derive new objects or to search for spatial relationships. Figure 2.8 shows a line object L_1 (solid lines) overlaid with another line L_2 (dashed lines).

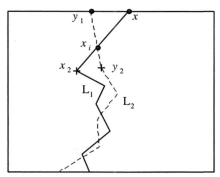

Figure 2.8 Intersection between two lines L_1 and L_2.

During the map overlay process all intersections between the two lines must be found. For example, line segments x_1 to x_2 and y_1 to y_2 will intersect if points y_1 and y_2 lie on opposite sides of the line from x_1 to x_2, and if x_1 and x_2 are on opposite sides of the line from y_1 to y_2. If the two line segments are shown to be intersecting, the point of intersection x_i is found by solving:

$$x_i = (1 - r_1)x_1 + r_1 x_2 = (1 - r_2)y_1 + r_2 y_2 \qquad (2.36)$$

where r_1 and r_2 are real numbers following the definition given in Equation 2.34.

Map overlay is an important GIS operation, and many useful spatial queries and analyses can be developed from it. For example, Figure 2.9(a) shows a buffering operation aimed at identifying those parts of line object L_1 lying within an expanded zone (grey-shaded) around a prescribed line (dashed lines). For area objects, map overlay may be usefully implemented to generate areas, such as A_1 and A_2 shown in Figure 2.9(b), which stand out in the light of certain given criteria.

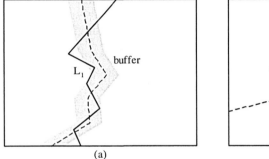 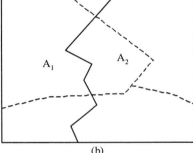

(a) (b)

Figure 2.9 Buffering (a) and polygonal overlay (b).

Objects have been treated as geometric primitives in this section so far. Object models may have imposed a view that position can be precisely measured, and in that sense are superior to fields, which suffer from finite resolution and spatial variation. But the high-precision coding and processing of objects in GIS has contributed to a false perception that accuracy requires little more than increasing the number of bits and sampling density. The problems associated with spurious precision have already been discussed, and a reassessment of object-based geo-processing is clearly helpful. If the two lines L_1 and L_2 shown in Figure 2.8 are actually different versions of the same unknown real line object, map overlay will become overwhelmingly complicated with many slivers, particularly when digitising precision is increased by densifying vertices for the polylines shown in Figure 2.6(b) (Goodchild, 1978).

With regard to spatial uncertainties in general terms, the processing of discrete objects is clearly more objective with finite resolution data structuring than with pure geometric notations of point coordinates and linear equations. If objects are encoded in spatial databases in a way that is adapted to their inherent uncertainties, slivers will cease to exist and will no longer hamper map overlay. For instance, if points and lines are represented as having finite sizes, as is suggested in the buffered zone shown in Figure 2.9(a), then objects of matching lineages will be seen as coinciding within quantifiable tolerances. In summary, if positional data are constrained not only to the boundaries of objects but also to some finite neighbourhood around the assumed true positions, then it is possible to view uncertain positional data as fields of point distributions. Fields are thus much better able to handle uncertain objects.

Spatial dependence has been discussed in Section 2.2.2 as being important for field variables. However, this should not be taken to suggest that spatial dependence is a phenomenon only applicable to fields. Spatial dependence is found to be inherent in objects as well, as was shown by Keefer *et al.* (1991), who studied error occurrence in the course of stream mode digitising of map points and lines. How to handle objects in the context of uncertainties and how to incorporate spatial correlation in object-based geo-processing remain topics of further work.

2.4 DISCUSSION

This chapter has described two geographical perspectives: fields and objects. Many aspects of landscapes are field-like, including elevation, surface roughness, and cover types. For instance, the elevation surface is continuous, and there is one value of elevation at an infinite number of possible places in geographical space. For field variables, there is an impressive body of theory and methods for describing and handling spatial variability and dependence, which are seen as crucial for understanding geographical phenomena. On the other hand, a point map of cities adopts an object view of the world, where the space between points is empty, and has no defined attributes. A street map is another object view of the world for which no attributes are defined between streets, since the world is empty except where there are streets with associated landmarks and traffic conditions. Discrete objects have a strong tradition in cartography and in map data processing, as well as in other forms of geographical information.

Map overlay is frequently referred to as a key function in GIS, which serves to combine information concerning objects and fields. In the domain of discrete objects, geometric operations are performed to identify intersections and to infer about topological relationships such as connectivity and containment. Object processing based on vector structures tends to be rather complicated due to the need for searching neighbourhoods and other spatial relationships in addition to performing geometric computing, which is itself intensive. In the realm of rasterised fields, on the other hand, algebraic and logic operators are easily implemented with spatially explicit evaluation of layered distributions. This comparison, commonly referred to in GIS, originates from the fundamental capability of field-based perspectives, as opposed to highly abstracted objects, in dealing with spatially varying distributions. In operational settings, the process of map overlay is often used to integrate a mixture of objects and fields. An example is relevant when one would like to derive statistics of ecotones for different areal units such as those of administrative and proprietary concerns. One should seek to combine data of heterogeneous nature efficiently and meaningfully, that is, in a way so that the strength of objects and fields is maximised in geographical computing but their integrity is maintained.

The widely acclaimed superiority of fields over discrete objects in geographical abstraction and uncertainty handling does not, in general, diminish the meaningfulness or usefulness of objects. Rather, objects, as a useful type of spatial data model, can be thought of as representing a perspective that is complementary to continuous fields, although examples such as land cover may be used to highlighting the benefits of using fields as opposed to objects for objectivity in representation and spatial analysis. Discussion of objects may benefit from a re-thinking in the light of fields and field-based methodologies. Likewise, understanding of field-like spatial distributions may sometimes benefit from the insights provided by an object perspective.

Consider a spatial query for variables at unsampled zones (polygons) based on data from source zones; such queries have been termed *areal interpolation* (Goodchild and Lam, 1980). For example, given population counts for census tracts, it is sometimes required to estimate populations for electoral or school districts. It would be easy if the target zones were aggregations of the source zones. But the boundaries of the two sets of zones are often independent of each other. One plausible approach is to calculate the population density for each source zone (e.g., a census tract) and to assign this value to the zone's centroid, a point object. Using this set of density data, a gridded population density surface may be derived using any reasonable method of spatial interpolation. Then, each grid cell's value is converted to a population by multiplying the estimated density by the cell's area. Finally, total populations in individual target zones are estimated by overlaying the grid of interpolated population on the target map.

However, this procedure is riddled with pitfalls, such as ill-defined centroids, inadequacy of point-based interpolators for areal problems, the uncertain behaviours of density surfaces outside the study area, and the possible presence of lakes and other unoccupied areas. The technique clearly makes use of both field and object perspectives, with the latter framing the spatial dimensions of the former. In general, the modification of field-based spatial analysis with discrete object perspectives has great potential. Useful methods are discussed by Rathbun (1998) for Kriging dissolved oxygen and salinity in an irregularly shaped estuary.

The complementarity between fields and objects is particularly helpful for handling indeterminate objects, whose uncertainty in the spatial and semantic domains is difficult to formulate. The interaction of positional and attribute uncertainties would be better explored within an integrated vision and approach. It is to the ultimate aim of building an integrated framework for handling geographical uncertainties that this chapter is dedicated.

Uncertainty seems to act to blur the cognitive and methodological boundaries between objects and fields. The duality of data models is perhaps a legacy of earlier perspectives on geographical phenomena, and less appropriate now that digital methods and technologies have enhanced the possibility of handling multi-faceted spatial uncertainties. It is logical to assert that certain unified strategies for representation and communication of geographical information may better be conceptualised and built upon encapsulation of uncertainty in geographical computing than on its separate handling. This vision will be expanded in Chapter 4, where uncertainty in both fields and objects is analysed on a spatially explicit basis, by reconciling the strengths and weaknesses of both probabilistic and fuzzy schools of thinking.

Conventional map-making and data processing are usually related to a single, often unspecified level of resolution or scale. The pervasive presence of uncertainty in geographical information suggests that it should be modelled as a fundamental component of the geographical world, and incorporated in GIS so that spatial query and decision-making are truly well informed, not so much because of the use of digital methods, but due to the potential for effective communication of uncertainty.

CHAPTER 3

Geographical Measurement and Data

3.1 INTRODUCTION

The value of geographical information for a specific purpose relies on the information's accuracy and fitness for that purpose as much as on the meaningfulness of data modelling and geo-processing. Access to geographical data is often quoted as the most fundamental requirement for GIS-based applications, overtaking hardware purchasing and software engineering in the hierarchy of issues related to information infrastructure (Goodchild, 1996). This assertion implies that data acquisition and production should not only be geared towards a broad range of purposes, but also should be accountable to applications in ways that extend far beyond traditional non-spatial notions concerning measurement.

In certain situations, geographical data may be obtained from secondary sources such as existing maps, tables, or other databases. In other situations, primary data may be derived from measurements conducted on the spot or via remote sensors. Awareness of data accuracy requires that assessment be made of the instruments and procedures used for measurement and data compilation, and also that measures of resolution, accuracy, and other indicators should be included in the data documentation, in other words should be recorded in metadata. Land or soil surveyors are trained to be concerned with inaccuracy when they perform measurement and sampling, and methods of data reduction, such as least squares, provide measures of accuracy in addition to the best estimates of true values.

In general, information about data inaccuracy may be obtained as a by-product of the process of data reduction, and can become a valuable addition to the data sets that are used to solve spatial problems. Data producers are likely to have detailed records or logs of the specifics of spatial sampling and map making, and these are clearly of profound value in assessing the accuracy of data. Such uncertainty-aware strategies can be built upon practical evaluation of the techniques commonly used for geographical mapping, together with statistical analyses of the effects of uncertainties during the data reduction process. Once information on data lineage is properly documented, geographical information can be communicated between data producers and users in ways that provide valuable insights concerning data sources and their inherent inaccuracies.

Land surveying using scientifically designed instruments plays an important role in data collection for modern GIS as well as in older forms of topographic mapping. It is used to derive the well-distributed ground control points (GCPs) that are required for proper geo-referencing of the separate data layers or contiguous data sets that are required in a particular GIS application. This function is particularly valuable in urban and suburban mapping, where many layers of data need to be co-registered for spatial analysis and planning. Global positioning systems (GPSs) have been developed since the 1980s and increasingly are replacing traditional land surveying systems as efficient ground data collection tools, with improved precision and reduced cost. The links between GPS and GIS are developing rapidly.

Ground-based surveying, especially when enabled by GPS, is vital for geo-referencing interpretations and measurements based on aerial photographs and remotely sensed images, as will be described in this chapter. While land surveying is able to provide accurate GCPs for many GIS applications, it would be impractical and prohibitive to employ land surveying to perform point-by-point detailed mapping in large areas, where efficiency and currency are often more crucial for the rapid production of current geographical information. This is where photogrammetric techniques can help.

Photogrammetry provides the technique for obtaining reliable information about physical objects and landscapes through the measurement and interpretation of photographic images, and patterns of electromagnetic radiant energy recorded as images. The photogrammetric stereo plotter is the instrument most commonly used to transfer aerial photography information to planimetric and topographical maps. In a stereo plotter, the operator views two mutually overlapping aerial photographs taken from different positions, to form a three-dimensional model. By moving a floating point around in the stereo-model, the operator can draw in roads, rivers, contours, and other entities. If the plotter has a suitable digital recording system, all movements used in drawing within the photogrammetric model can be numerically encoded and entered into computer storage. Recent developments in photogrammetry have significantly improved its efficiency and currency in spatial data acquisition and have provided increased accuracy. This is especially apparent in digital or *soft* photogrammetry, which proceeds through the automatic correlation of digital images, and offers great opportunities for automated feature identification and image interpretation, as well as for geometric measurement.

An extension of aerial photo-interpretation in a sense, remote sensing has played increasingly important roles in the provision of geographical data. Remote sensors may operate on airborne or spaceborne platforms. An example of the former is the airborne thematic mapper (ATM) sensor on board aeroplanes, providing data with fine resolution (pixel sizes ranging from sub-metre to several metres). On the other hand, spaceborne remote sensors on a number of satellites in orbit, such as Landsat MSS (multispectral scanner), Landsat TM (thematic mapper), and SPOT HRV (high resolution visible) are providing timely and regular coverages that are useful for deriving data about natural resources and the environment from regional to global scales.

In some situations, remote sensing data are used to provide direct estimates of certain physical variables independently of other data. Operational examples are found in studies related to vegetation or forestry, where the relevant spatial resolution and extent are frequently comparable to those of the satellite data, as argued by Harris (1987). Although urban land cover cannot be divided readily into discrete classes with conventional image classifiers, satellite sensors with a relatively fine spatial resolution, such as SPOT HRV or Landsat TM, have considerable potential for data updating and monitoring, and for studies of such dynamic geographical phenomena as urban sprawl (Weber, 1994). Certain variables may be inferred from the measurement of related variables from remotely sensed data. For example, by relating population density to building characteristics such as building type and coverage, remotely sensed data may be used to derive estimates of demographic variables. Successful experiments along these lines were carried out by Langford and Unwin (1994) using classified Landsat TM data to derive population density surfaces based on estimates of built-up areas.

This chapter will review some of the mainstream techniques for spatial data acquisition, highlighting the characteristics of errors and uncertainties that occur in spatial measurement and mapping. It is hoped that, by describing some of the technical details involved in data acquisition, it may become easier to identify and analyse errors and their occurrences. Producers and users of geographical data can, therefore, reach a well-informed consensus about tolerable and predictable errors, which might otherwise go unnoticed, with unpredictable results.

3.2 LAND SURVEYING AND POSITION-FIXING

Traditional land surveying requires skill and experience in the observation of angles, distances, and height differences. The introduction of the electronic theodolite or total station (i.e., an electronic theodolite integrated with or attached to an electronic distance-measuring device) has made conventional surveying easier and faster. Some established procedures are made available as standard programmes and are even built into GIS packages as sub-systems (e.g., the coordinate geometry or *COGO* commands in ARC/INFO), thus computerising the entry of angle and distance measurements and subsequent coordinate calculation.

The availability of sophisticated instruments and the refinement of surveying procedures have led to efficient treatment of the errors inherent in the surveying process. Most importantly, land surveyors check the accuracy of their work directly in the field using highly refined surveying procedures or rules. Also, by survey adjustment techniques (Cooper and Cross, 1988; Wolf and Ghilani, 1997), field measurements can be modified very precisely into most probable values. Thus, it is appropriate to say that uncertainties are relatively easy to measure and control in land surveying. The understanding and handling of errors will become more important as the automation of surveying work generates increasing amounts of data, whose errors may accumulate and propagate very quickly, especially when land surveying is carried out over large areas with a high density of detail.

Surveyors make measurements with different types of instruments, which are used to derive other quantities, such as three-dimensional Cartesian coordinates for precision positioning. A functional model exists that can be used to relate measured quantities and their associated errors with these derived quantities.

Consider the surveying network shown in Figure 3.1, where the coordinates of five new points x_2 to x_6 are to be determined from angle and distance measurements in a coordinate system defined by the given eastings and northings of the two known control stations x_1 and x_7.

Denote the planimetric coordinates of two points x_{s1} and x_{s2} as (X_{s1}, Y_{s1}) and (X_{s2}, Y_{s2}) respectively. Planimetric coordinates are related to the distance measurement $dis(x_{s1}, x_{s2})$ between x_{s1} and x_{s2} via a functional model (Cooper and Cross, 1988):

$$dis(x_{s1}, x_{s2}) = \|x_{s1} - x_{s2}\| = \left((X_{s1} - X_{s2})^2 + (Y_{s1} - Y_{s2})^2 \right)^{1/2} \tag{3.1}$$

On the other hand, each angular measurement involves three points. For point x_s at the angle corner, denote its backward and forward points as x_{s-1} and x_{s+1}.

The angular measurement $\theta(x_{s-1},x_s,x_{s+1})$ corresponds to an observation equation as follows:

$$\theta\left(x_{s-1},x_s,x_{s+1}\right) = azimuth\left(x_{s-1}-x_s\right) - azimuth\left(x_{s+1}-x_s\right)$$
$$= \arctan\left(\left(Y_{s+1}-Y_s\right)/\left(X_{s+1}-X_s\right)\right) - \arctan\left(\left(Y_{s-1}-Y_s\right)/\left(X_{s-1}-X_s\right)\right) \qquad (3.2)$$

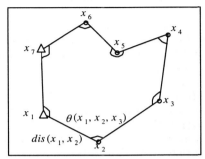

Figure 3.1 A surveying network for horizontal control.

Levelling is undertaken to establish vertical control. Suppose a network of levelling is designed as shown in Figure 3.2, where traversal directions are marked by arrows. Suppose two benchmarks x_1 and x_3 are available with Z_1 and Z_3 being the known heights, and that height differences $lev(x_1,x_2)$ and $lev(x_2,x_3)$ are measured between point pairs x_1 and x_2, and x_2 and x_3, respectively The unknown height Z_2 for point x_2 is implicitly related to the two benchmarks as follows:

$$lev\left(x_1,x_2\right) + lev\left(x_2,x_3\right) = \left(-Z_1+Z_2\right)+\left(-Z_2+Z_3\right) = -Z_1+Z_3 \qquad (3.3)$$

where the unknown value Z_2 no longer appears. In this case, a conditional equation rather than two observation equations may be formulated as above.

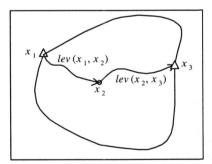

Figure 3.2 A levelling network for height control.

Observational and conditional equations, such as those in Equations 3.1, 3.2, and 3.3, are called functional models, and can be represented generically as:

$$f\left(UU,SU,con\right) = 0 \qquad (3.4)$$

where UU is the vector of elements to be evaluated (e.g., coordinates), SU is the vector of elements that have been measured (e.g., distances and angles) and con is the vector of elements that are known as constant or regarded as so.

Functional models are true by the axioms and logic of Euclidean space and within the abstract space of that geometry. However, if functional models are chosen to represent relationships in an empirical sense, they are no longer true. The effects of the atmosphere, state of the theodolite, incorrect pointing, and other physical factors render functional models inaccurate, leading to systematic errors when considered in connection with a particular measurement and functional model.

It is not possible to eliminate systematic errors, but their effects can be reduced by devising suitable methods for calibrating instruments, and by applying numerical corrections to measurements, or by refining measurement procedures, such as the principle of reversal, so that specific physical effects can be reduced or even cancelled out. Moreover, certain physical effects may be represented as additional terms in the functional model of Equation 3.4. Examples include the scaling errors in distance measurements obtained with a tape or a ruler.

Another type of error, which has nothing to do with a functional model, is a gross error, also known as a blunder or mistake. Generally, a gross error arises from an incorrect measuring or recording procedure, or an error in a computer program. A gross error is easily detected and rectified by independent checks.

Even if systematic and gross errors were eliminated, there would still be errors in measurements, since it is known that repeated measurements of the same element, made in quick succession and perhaps with the same instrument, would appear inconsistent. These inconsistencies cannot be removed by refining the functional model or by applying corrections. They are often caused by random errors, whose magnitude and sign are subject to stochastic events. It is thus customary to regard measurements as random variables and to describe them using stochastic models.

Expectation and variance, which are parameters of the underlying population, are useful descriptors of a random variable that is normally distributed. In practice, with a sample of that population (e.g., su_i, $i = 1,2,\ldots,n$, as a sample of size n for a variable su), it is possible to obtain unbiassed estimates for expectation and variance as:

$$m_{su} = \sum_{i=1}^{n} su_i / n \tag{3.5}$$

and:

$$\text{var}_{su} = \sum_{i=1}^{n} (su_i - m_{su})^2 \Big/ (n-1) \tag{3.6}$$

where m_{su} and var_{su} are the mean and variance estimated from a sample of measurements (with the latter being the squared form of standard deviation).

When a pair of random variables, such as $su1$ and $su2$, is considered, covariance is an important descriptor for the correlation between them. An unbiassed estimate of the covariance can be computed from a sample of measured data by:

$$\text{cov}(su1, su2) = \sum_{i=1}^{n} (su1_i - m_{su1})(su2_i - m_{su2})/(n-1) \tag{3.7}$$

where m_{su1} and m_{su2} stand for the means of variables $su1$ and $su2$, respectively, and n is the sample size of the measurements.

Given a set of measured data SU and knowledge of their variance and covariance matrix COV_{SU}, unknown elements UU can be evaluated using a specific function model. But, this may not be straightforward due to the non-linearity of many functional models and the stochastic nature of measured data, although, for the levelling network designed in Figure 3.2, a linear functional model is readily available. Nevertheless, by linearisation and least squares, most probable estimates of unknowns can be derived from measurements constrained by certain functional models.

By applying Taylor's theorem, Equation 3.3 can be approximated to first order as:

$$f(UU, SU, con) =$$

$$f(UU', SU', con') + (UU^* - UU')\partial f/\partial UU + (SU^* - SU')\partial f/\partial SU =$$

$$f_0 + M_A d_{UU} + M_B V_{SU} = 0 \tag{3.8}$$

where adjusted values for UU and SU are denoted by UU^* and SU^*, while UU' and SU' represent the first-order approximations for them, at which the partial differentials are evaluated. In the third line of Equation 3.8, f_0 represents the first-order approximation of the functional model f, M_A and M_B are matrices of the partial differential coefficients, and d_{UU} and V_{SU} are the corrections to UU' and SU', respectively. If it is assumed that SU' corresponds to the measured values SU, then the term $SU^* - SU'$ represents the corrections V_{SU} to the measurements SU to give the adjusted values SU^*.

There is usually redundancy in measurement, because the numbers of measured values and constraints, such as Equation 3.3, are often more than enough to determine the number of unknowns. A unique solution to the linearised functional model in Equation 3.8 for first-order approximations d_{UU} and V_{SU} can, however, be obtained by requiring the quadratic form $V_{SU}^T COV_{SU}^{-1} V_{SU}$ be a minimum as:

$$d_{UU} = -\left(M_A \left(M_B COV_{SU} M_B^T \right)^{-1} M_A \right)^{-1} M_A^T \left(M_B COV_{SU} M_B^T \right)^{-1} f_0 \tag{3.9}$$

and:

$$V_{SU} = -COV_{SU} M_B^T \left(M_B COV_{SU} M_B^T \right)^{-1} \left(M_A UU^* + f_0 \right) \tag{3.10}$$

As it is generally not possible to obtain values UU' to first-order accuracy, the solution for UU^* has to be iterative. The adjusted values UU^* thus become the approximation for the next round of iteration, until the corrections d_{UU} become insignificantly small. Not only is it important to obtain estimates for the unknowns and adjustments for measurements, but their variances and covariances are important also. Application of the law for the propagation of covariance leads to an evaluation of the covariance matrix for the derived unknowns. For instance, let COV_{UU*} be the covariance matrix of the unknowns, derived from least squares as $UU' + d_{UU}$. This matrix is computed from:

$$COV_{UU*} = \left(M_A^T \left(M_B COV_{SU} M_B^T \right)^{-1} M_A \right)^{-1} \tag{3.11}$$

Now that the method of least squares has been described, it is possible to deal with the topic of data reduction in land surveying. Equation 3.1 is approximated by a first-order Taylor expansion of the following form:

$$dis(x_{s1}, x_{s2}) - \|x_{s1}' - x_{s2}'\| + \|x_{s1}' - x_{s2}'\|^{-1}((X_{s2}' - X_{s1}')(dX_{s1} - dX_{s2})$$

$$+ (Y_{s2}' - Y_{s1}')(dY_{s1} - dY_{s2})) + v_{dis(x_{s1}, x_{s2})} = 0, \tag{3.12}$$

where $\|x_{s1}' - x_{s2}'\|$ is the approximate Euclidean distance calculated from the initial coordinates (X_{s1}', Y_{s1}') and (X_{s2}', Y_{s2}') for points x_{s1} and x_{s2}, whose increments are denoted by dX_{s1}, dY_{s1}, dX_{s2}, and dY_{s2}, respectively. In linearisation, known points are commonly taken as constants, and are not to be subjected to adjustment. Thus, dX_1 and dY_1 will not show up in the linearised functional model for $dis(x_1, x_2)$, since point x_1 is known, as shown in Figure 3.1.

Following the same approach, Equation 3.2 is linearised to:

$$\theta(x_{s-1}, x_s, x_{s+1}) - (azimuth(x_{s-1}' - x_s') - azimuth(x_s' - x_{s+1}')) -$$

$$((Y_s' - Y_{s-1}')/\|x_{s-1}' - x_s'\|^2)dX_{s-1} + ((X_s' - X_{s-1}')/\|x_{s-1}' - x_s'\|^2)dY_{s-1} +$$

$$(((Y_s' - Y_{s-1}')/\|x_{s-1}' - x_s'\|^2) + ((Y_{s+1}' - Y_s')/\|x_s' - x_{s+1}'\|^2))dX_s -$$

$$(((X_s' - X_{s-1}')/\|x_{s-1}' - x_s'\|^2) + ((X_{s+1}' - X_s')/\|x_s' - x_{s+1}'\|^2))dY_s -$$

$$((Y_{s+1}' - Y_s')/\|x_s' - x_{s+1}'\|^2)dX_{s+1} + ((X_{s+1}' - X_s')/\|x_s' - x_{s+1}'\|^2)dY_{s+1} +$$

$$v_{\theta(x_{s-1}, x_s, x_{s+1})} = 0 \tag{3.13}$$

where $azimuth(x_{s-1}' - x_s')$ and $azimuth(x_s' - x_{s+1}')$ are calculated using initial values of coordinates (X_{s-1}', Y_{s-1}'), (X_s', Y_s'), and (X_{s+1}', Y_{s+1}') for points x_{s-1}, x_s, and x_{s+1}, whose increments are denoted by dX_{s-1}, dY_{s-1}, dX_s, dY_s, dX_{s+1}, and dY_{s+1}, respectively.

Each distance measurement establishes one observation. Thus, for the example shown in Figure 3.1, there will be six observation equations based on distance measurements, while angular measurements lead to five observation equations. In total, there will be eleven observation equations over-determining ten coordinate values for five unknown points, leaving one redundancy.

By rigorously implementing least squares for a surveying network, it is possible to obtain estimates of three-dimensional coordinates with high accuracy without major difficulties. This is based on the fact that many measurement procedures are equipped with angle- and distance-measuring instruments with standard errors of 1.0 second and 4.00 ppm (parts per million), while levelling accuracy is commonly within 1 centimetre (Cooper and Cross, 1988).

Thanks to developments in computerised coordinate determination in land surveying, surveying networks of distance and angle measurement anchored to triangulated stations of known coordinates can be easily solved for unknown coordinates. In such processing, accuracy is also evaluated, most commonly by applying the law of variance propagation, as has been described previously in this section. Thus, it seems that ground-surveyed data have secured adequate quality by rigorous computation and analysis.

A major advancement in surveying is the operational use of GPS, which is built upon a system of geostationary satellites in orbit around the Earth, whose locations are predictably and accurately known. Thus, a receiver capable of decoding signals from these satellites will be able to provide positional data by three-dimensional triangulation from a set of three satellites. Further, on a mobile platform coupled with microcomputer postprocessing, it is not difficult to obtain real-time geo-referencing in terms of latitude, longitude, and elevation. GPS-based positional data can be precise within about one centimetre, given sophisticated procedures for data reduction (Clarke, 1995; Kaplan, 1996). Therefore, GPS technologies have found increasing applicability for surveying remote mountains, deserts, and islands.

It is clear that applying ground-based surveying for precision positioning of objects is not economical for detailed mapping over large areas. GPS-based surveying may also not be easily accessible or practical for mapping geographical distributions in detail. Photogrammetric and remote sensing techniques are more suited for mapping geographical detail by providing comprehensive coverage, cost-effectiveness, and a long tradition of topographical mapping, as the following sections seek to establish.

3.3 PHOTOGRAMMETRIC MAPPING

Photogrammetry, which began as a technique for reconnaissance, has matured into a distinctive discipline and has been consolidated as a large industry for image-based geographical information. Aerial photographs are typically the sources from which photogrammetrists seek to extract useful information for creating maps detailing topography and various geographical themes.

An image point on an aerial photograph, which is actually of finite granularity, is positioned using a coordinate system *xoy* in a two-dimensional image plane that is defined with fiducial marks, such as the tiny triangles and circles in Figure 3.3. Data about fiducial marks are usually calibrated and supplied by manufacturers of aerial survey cameras. From a comparator with magnification, it will be possible to measure coordinates for an image point j as (x_j, y_j), as is illustrated in Figure 3.3, to a precision of a few micrometers (Trinder, 1973). These image measurements provide the sources to derive land information that is geo-referenced to a chosen ground coordinate system, thus playing roles that are comparable to those played by distance and angle measurement on the ground.

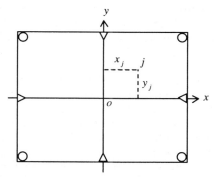

Figure 3.3 Coordinate system for an aerial photograph.

In addition to fiducial marks, a focal length or camera constant is also part of the calibration data. Specification of a principal point with reference to the origin of the image coordinate system *xoy*, along with camera focal length, enables a precise modelling of the bundle of light rays at the instant of photographic exposure. Thus, aerial photographs record images of landscapes according to perspective projection, as is shown in Figure 3.4, where a point J on the ground is imaged as j on the photograph. For an aerial photograph that is flown at a height H above a reference plane and taken normal to the surface using a camera of focal length f, nominal scale is defined as the ratio of focal length to the camera height; in other words, the photographic scale is f/H.

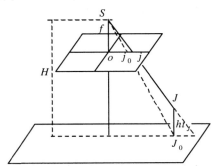

Figure 3.4 Perspective projection in an aerial photograph.

However, it is not accurate to use this nominal scale to recover the actual distance between two points on the ground from aerial photographs, unless the terrain is flat without undulation. As is shown in Figure 3.4, the ground point J with a relief ht_J above the reference plane containing its ortho-projection point J_0 should be imaged as j_0 on the photographic plane to recover its correct ortho-projection position, leaving a displacement vector $j - j_0$. It is easy to calculate the photographic displacement due to a relief of ht_J at point J from the equation:

$$\|j - j_0\| = \|j - o\| ht_J / H \qquad (3.14)$$

where $\| j\text{–}o \|$ stands for the radial distance of the image point j from the principal point o.

With respect to the projective geometry shown in Figure 3.4, one can clearly see that with a given aerial photograph flown at specific height, tall buildings and trees lying towards the photograph edge will be imaged to lean outwards. Moreover, the orientation of imaged objects is generally distorted, precluding aerial photographs from being used as ortho-projection maps.

To reconstitute rigorously the bundle of light rays at the instant of exposure, in addition to the interior orientation elements including fiducial marks and the camera constant, six further elements are required for exterior orientation. As shown in Figure 3.5, an image space coordinate system is defined as xyz centred at the camera station S distanced by focal length f from the image plane, where Sx and Sy are parallel to the image coordinate axes ox and oy shown already in Figure 3.3. The camera station S is located in a ground coordinate system XYZ with coordinate (X_S, Y_S, Z_S), with the bundle of exposure rays tilted at angles of φ, ω, and κ, as illustrated in Figure 3.5.

These three angular elements determine the amounts that are needed to rotate the bundle of rays from the ideal vertical direction to the actual tilted attitude of exposure. This is simulated by firstly rotating through an angle of φ around the Y

axis, secondly through an angle of ω around the then-rotated X axis, and finally through an angle of κ around the then-rotated Z axis.

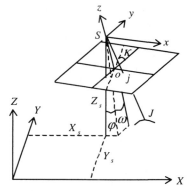

Figure 3.5 Orienting an aerial photograph.

An important geometric relationship in photogrammetry is the so-called co-linearity equation, which can be established from the orientation elements mentioned above. By the nature of perspective projection, the image point j and its corresponding ground point J are co-linear. Thus, the two vectors, $J\text{–}S$ and $j\text{–}S$ differ only in magnitude but coincide in direction, that is:

$$J - S = mg_J (j - S) \tag{3.15}$$

where mg_J stands for a scale factor or denominator at point J.

In order to express the vector relationship stipulated in Equation 3.15 as coordinates, the image vector $j\text{–}S$ has to be referenced with respect to the ground coordinate system XYZ. This orientation is quantified as a rotation matrix $R_{\varphi\omega\kappa}$, which is the product of a sequence of three rotation matrices and may be evaluated as the following matrix:

$$R_{\varphi\omega\kappa} = \begin{pmatrix} b_{11} & b_{12} & b_{13} \\ b_{21} & b_{22} & b_{23} \\ b_{31} & b_{32} & b_{33} \end{pmatrix} \tag{3.16}$$

where matrix elements b_{11}–b_{33} are all functions of the three angular parameters φ, ω, and κ.

Based on the rotation matrix in Equation 3.16, it is possible to express the vector equation in coordinate components. By removing the scale factor mg_J in Equation 3.15, it is possible to arrive at the following co-linearity or photographic imaging equation:

$$(X_J - X_S)/(Z_J - Z_S) = (b_{11}x_j + b_{12}y_j - b_{13}f)/(b_{31}x_j + b_{32}y_j - b_{33}f)$$

and:

$$(Y_J - Y_S)/(Z_J - Z_S) = (b_{21}x_j + b_{22}y_j - b_{23}f)/(b_{31}x_j + b_{32}y_j - b_{33}f) \tag{3.17}$$

However, the co-linearity in the photographic system expressed by Equation 3.17 says little about the Z dimension, as the viewing depth of point J can be anything along the perspective vector SJ. One of the most important photogrammetric principles is stereoscopy, which exploits parallaxes from two overlapping aerial photographs to detect differences in viewing depth, and hence to reconstitute three-dimensional geometry in the object space *XYZ*.

An illustration of an ideal photographic stereo-pair is provided in Figure 3.6, where two overlapping aerial photographs imaging object point *J* as *j* and *j'* are exposed at a pair of camera stations *S* and *S'* forming the baseline *SS'*. As shown in Figure 3.6, the two image-space coordinate systems *S-xyz* and *S '-x'y'z'* are defined parallel to their respective image plane coordinate axes, crossing principal points *o* and *o'*, which are also parallel to the object-space coordinate system *XYZ*.

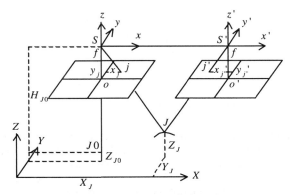

Figure 3.6 Mapping from a reconstituted photogrammetric stereo-model.

Suppose that the homologous image points *j* and *j'* are measured with coordinates (x_j,y_j) and $(x_{j'},y_{j'})$. For the two rays projected through *j* from *S* and through *j'* from *S'* to intersect, y_j and $y_{j'}$ must be equal, leaving zero *y*-parallax (*y*-parallax=y_j–$y_{j'}$=0), for stereo viewing.

On the other hand, the *x*-parallax, px_j=x_j–$x_{j'}$, is useful for discerning the difference of points in the object space from the perspective centres *S* and *S'*. Refer to the geometry implied in Figure 3.6. It is possible to calculate the vertical distance or height H_J of camera *S* from point *J* as:

$$H_J = \left\|S' - S\right\| f /\left(x_j - x_{j'}\right) = \left\|S' - S\right\| f / px_j \qquad (3.18)$$

from which it is further possible to calculate:

$$X_J = H_J x_j / f$$

and:

$$Y_J = H_J y_j / f \qquad (3.19)$$

Suppose there is another point *J0* of known *Z*-coordinate Z_{J0} with reference to a certain datum. It is easy to formulate the camera height with respect to point *J0*, H_{J0}, using Equation 3.18. Thus, the height difference between points *J* and *J0*, $H_{J\text{-}J0}$ =H_{J0}–H_J, is calculated as:

$$H_{J-J0} = \left\|S' - S\right\| f \left(px_j - px_{j0}\right) /\left(px_j px_{j0}\right) = H_{J0}\, dpx_j /\left(px_{j0} + dpx_j\right) \qquad (3.20)$$

where dpx_j is the so-called *x*-parallax difference between an unknown point *J* and a reference point *J0*. From Equation 3.20, it is straightforward to calculate the *Z* coordinate of point *J* as the sum of Z_{J0} and $H_{J\text{-}J0}$.

However, the description above is based on an idealised situation where two overlapping aerial photographs are flown vertically facing down and at a constant flight height. This is hardly close to a practical scenario, in which aerial photographs are imaged subject to varying attitudes and flight instability, as shown in Figure 3.7, highlighting the complexity of photogrammetric processing.

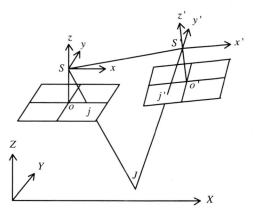

Figure 3.7 Relative and absolute orientations of an image pair.

In order to map the three-dimensional real world from overlapping aerial photographs, the image pair shown in Figure 3.7 must be adjusted through a combination of relative and absolute orientation, which are jointly known as exterior orientation, as opposed to the interior orientation that is facilitated through calibration using fiducial marks and the camera constant. In a relative orientation, the left photograph is usually assumed to be fixed, with the right photograph adjusted to recover relative camera position and exposure attitude. To obtain a rigorous solution to relative orientation, homologous rays have to intersect in a plane called the epipolar plane containing the points S, S', j, j', and J as shown in Figure 3.7. A conditional equation for relative orientation is known as the co-planarity condition:

$$(S'-S)\cdot((J-S)\times(J-S'))=0 \qquad (3.21)$$

where the symbols $\cdot$ and $\times$ represent dot and vector products respectively. One such condition equation can be written for each pair of homologous image points. Least squares is then applied for solving the parameters implied in Equation 3.21 for relative orientation. This is not further elaborated here.

Upon accomplishing the relative orientation of an image pair as above, the stereo-model must be referred to a ground coordinate system XYZ to enable the positioning of objects therein, and the production of maps at a chosen scale. This process is known as absolute orientation, in which model coordinates for points such as J, which are often referenced to S-xyz, are transformed into the ground coordinate system XYZ as follows:

$$(XYZ)^{\mathrm{T}}_{J} = (XYZ)^{\mathrm{T}}_{J0} + mgR((xyz)_{J} - (xyz)_{J0})^{\mathrm{T}} \qquad (3.22)$$

where $(XYZ)^{\mathrm{T}}_{J0}$ and $(xyz)^{\mathrm{T}}_{J0}$ are coordinates in object and model spaces for a known point $J0$, while mg stands for a scale factor and R for the rotation matrix required for a linear transform of coordinates from model to object spaces.

Absolute orientation is carried out based on the availability of a certain number of ground points of known coordinates. Both relative and absolute orientations are solved iteratively by linearisation of their respective functional models, that is, Equations 3.21 and 3.22.

Linearisation of photogrammetric functional models involves perspective projection via rotations between coordinate systems. For aerial photography with minor tilts of $d\varphi$, $d\omega$, and $d\kappa$, Equation 3.16 may be approximated with first order accuracy as:

$$dR_{\varphi\omega\kappa} \approx \begin{pmatrix} 1 & -d\kappa & -d\varphi \\ d\kappa & 1 & -d\omega \\ d\varphi & d\omega & 1 \end{pmatrix} \tag{3.23}$$

where d prefixes stand for differentials of the corresponding quantities. This approximation is useful, especially, for photogrammetric mapping based on analogue plotters using mechanical and optical devices to simulate perspective projection, and for iterative analytical data reduction aided with computers. The latter case can make use of the least squares approach discussed in the previous section for solution of the orientation parameters and evaluation of variance in photogrammetric coordinate determination.

The accuracy obtainable with photogrammetric positioning has been analysed in detail by Hallert (1960). Planimetric and height accuracies were quantified using variance propagation with reference to six hypothetical GCPs for an image pair, which were located at standard positions, that is, in the vicinity of principal points and at fixed y distances from them. It was found that the standard error in planimetry increases outward from the centre of the overlapping area of the stereo-model to the edges. At the model centre, the standard error at photographic scale is about 1.2 times the standard errors in measurements of both x and y image coordinates, while a factor of 2.0 is estimated near model edges. The overall pattern of planimetric error could be nicely viewed as a continuous surface quantifying the level of accuracy with which every image point is potentially positioned.

The standard error in height estimation was examined similarly via variance propagation and mapped as a surface. The surface of standard error in photogrammetric heights was contoured as near-concentric ellipses with the major axis directed along the y image axis. As can be anticipated, the surface of standard error in photogrammetric heights grows outwards from the vicinity of the model centre at about 2.0 times the standard error in measuring x-parallax, to a factor of about 3.0 at the edges (Hallert, 1960).

An inference that can be drawn from this analytical study is that, given quality photographs, precision plotters, careful stereoscopic viewing, and acuity by photogrammetrists, it is possible to achieve accuracies of several centimetres in photogrammetric mapping. With the development of analytical photogrammetry featuring numerical solutions of internal and external orientations, aerial-triangulation was provided with rigorous least squares approaches to the solution of photogrammetric block adjustment, from which photo-control is densified from a few known ground points and their photographic measurement (Grün, 1982). An elaboration is, however, outside the scope of this book.

For the majority of topographic and thematic mapping projects, photogrammetric reduction of source images renders a point-by-point treatment of errors in measurement prohibitively expensive. Besides, many spatial problems involve interpretative processing rather than merely positioning, making a pure analytical examination of errors in photogrammetric processing hardly, if ever, possible. The remainder of this section will document some practical aspects of photogrammetric mapping and some of the principal techniques of digital photogrammetry, and will focus on some of the issues related to automation and accuracy in photogrammetry.

It was noted previously that a photographic stereo-model is reconstituted through interior orientation with a principal point and focal length, and exterior orientation of a photographic pair with respect to a specific ground coordinate system. Upon completion of stereo-model reconstitution, photogrammetric mapping is performed to delineate relevant entities and to place them on maps as points, lines, and areas. While spatial data are recorded in analogue plotters in a similar way to the manual drafting of paper maps, with analytical plotters, any objects that are tracked using the measuring mark by a photogrammetrist are instantly located with coordinates and can be encoded digitally. This feature enables the easy creation of spatial databases with enhanced currency and with more flexible object selection than is possible in digitising from paper maps.

Another important type of data produced from photogrammetry is the digital elevation model (DEM), which is generated in a highly cost-effective way, especially with analytical plotters (Torlegård *et al.*, 1986). Spot heights may be measured at irregularly spaced points, or profiles of terrain can be sampled at regular intervals. This primary source for creating DEMs is frequently noted as an advantageous aspect of photogrammetry as opposed to sampling from existing contour maps, since the former is not subject to the types of generalisation and approximation that are common in the creation of topographic maps.

Many of the recent developments in photogrammetry are related to digital or softcopy photogrammetry, in which digital images are input to various photogrammetric procedures running on computers (Schenk, 1994; Greve, 1996). Instead of photographic hardcopies being placed on analogue or analytical plotters with varying combinations of mechanical, optical, and electronic components, images digitised by scanning or acquired directly in digital form are input to photogrammetric treatment. A digital plotter is actually a computer that performs mathematical processing of stereo-model reconstitution and planimetric mapping and height estimation with increasing degrees of automation.

One of the major differences between photographic and digital images is related to spatial resolution. For a photographic image, spatial resolution is expressed by the minimum ground width of identifiable black and white line pairs, while the diameter of ground coverage of each pixel defines the spatial resolution of a digital image. Scanning with an aperture of 10 μm yields a pixel size of 0.1 m on the ground for a 1:10,000 scale image. For a 1:24,000 scale aerial image scanned through an aperture of 40 μm, the corresponding ground resolution is about 1 m.

Digital photogrammetry opens the possibility for automatic production of digital maps. One of the main techniques in digital photogrammetry is image correlation, which seeks to replace human stereoscopic viewing by computer procedures that are capable of accurately matching homologous image points between left and right images. This process requires a search on the basis of an array of grey levels surrounding a specific point on the left image, for a corresponding point within a similar array of grey levels on the right image.

The general approach is to compare an array in the target area with arrays of the same size in the search area in order to find a match (Helava, 1978). Image matching may be carried in a two-dimensional way, as is shown by square segments on the left/right image pair in Figure 3.8. Two-dimensional image matching may be reduced to an easier one-dimensional routine. In this way, search is carried out along corresponding epipolar lines of the left and right images, as is illustrated by the tiny slits of image segments in Figure 3.8.

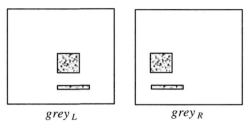

$$grey_L \qquad\qquad grey_R$$

Figure 3.8 Correlating an image pair.

A usual criterion is that successfully matched segments should register the strongest similarity in distributions of grey levels. Many useful methods have been devised for image matching that seek to maximise a specified correlation function. The classical correlation function for one-dimensional signals is denoted by $cor_{Z1Z2}(\tau)$ at a lag of τ:

$$cor_{Z1Z2}(\tau) = \frac{1}{T} \int_0^T Z1(t)Z2(t+\tau)dt \qquad (3.24)$$

where $Z1$ and $Z2$ stand for two signals within a period of T. In this equation, $cor_{Z1Z2}(\tau)$ is written as a continuous time-series function. For segments of grey levels $grey_L(x_j)$ and $grey_R(x_j)$ for a left and right image pair, the correlation function may be written in discrete form as:

$$cor_{Z1Z2}(dx) = (1/n)\sum_{j=1}^{n} grey_L(x_j)grey_R(x_j + dx) \qquad (3.25)$$

where two image segments of the same length of n pixels are displaced at a lag of dx. Other useful definitions of correlation strength in grey levels include those based on covariances and correlation coefficients. Interestingly, grey-level differences of neighbouring pixels may be usefully substituted for the original grey levels in the search for maximum correlation (Wang, Z., 1990).

To enhance the accuracy of image matching and its assessment, a least-squares approach can be implemented. Consider distributions of grey levels for target and search areas on the left and right images, which are represented as $grey_L(x_j)$ and $grey_R(x_j)$, respectively, with subscript j denoting the pixels within the image segments of identical size. For a one-dimensional implementation of image correlation, pixels x_j may be indexed with integers from 1 to n, for a total length of image segment of n pixels.

It is reasonable to conceive of actual grey levels imaged on aerial photographs as contaminated versions of ideal images $g0_L(x_j)$ and $g0_R(x_j)$. The errors in these contaminated versions are assumed to be noises $noise_L(x_j)$ and $noise_R(x_j)$ of zero means. Thus, it is possible to write an observation equation for the difference of grey levels as:

$$dgrey(x_j) = grey_R(x_j) - grey_L(x_j)$$

$$= g0_R(x_j) + noise_R(x_j) - (g0_L(x_j) + noise_L(x_j)), j = 1, ..., n \qquad (3.26)$$

For ideal imaging, grey levels $g0_L(x_j)$ and $g0_R(x_j)$ should be the same at a displacement of dx_j in a left image pixel x_j, that is, $g0_R(x_j) = g0_L(x_j - dx_j)$. Linearisation of Equation 3.25 is obtained by approximating $g0_L(x_j - dx_j)$ to first order and noting the assumption of uniformity of ideal grey levels at corresponding image points. Thus, a linearised form for the observation of difference in grey levels at pixel x_j is written as:

$$dgrey(x_j) = -gradient(g0_L(x_j))dx_j - (noise_L(x_j) - noise_R(x_j))$$

$$= -gradient(g0_L(x_j))dx_j - v(x_j) \qquad (3.27)$$

where $v(x_j) = noise_L(x_j) - noise_R(x_j)$ is the correction term for the observed difference in grey levels, that is, $dgrey(x_j)$.

Referring to Equations 3.9 and 3.11, it possible to arrive at an evaluation of the unknown dx_j and its variance, using a simplifying assumption that grey levels are equal-weight observations. This leads to:

$$dx_j = sum(gradient(g0_L(x_j))dgrey(x_j)) / sum((gradient(g0_L(x_j)))^2) \qquad (3.28)$$

and:

$$var(dx_j) = \left(\frac{2}{n}\right) var(noise) / var(gradient(g0_L))$$

$$= \left(\frac{2}{n}\right) var(grey) / (SNR^2 \, var(gradient(g0_L))) \qquad (3.29)$$

where SNR is the signal-to-noise ratio of the grey levels, and the variance of grey level gradient is clearly related to image texture (Wang, Z., 1990). A direct interpretation is that increased SNR, distinctive image texture, and an appropriately large size of target segment implied by n permit more accuracy in image matching.

Solutions for the unknown dx_j are obtained through an iteration process. Initially, $g0_R(x_j)$ and $g0_L(x_j)$ are set to be equal with zero displacement. After each iteration, grey levels are resampled from the non-zero dx_j and evaluated by least squares. Finally, it is possible to arrive at an adjusted value for dx_j, which concludes the derivation of x-parallax for one-dimensional image matching.

Successful matching of homologous image points leads to the automation of stereoscopic modelling of imaged terrain, given interior and exterior orientations. Thus, a dense array of points can be determined with ground coordinates. This suggests a possibly automatic production of DEMs over regular grids, which holds considerable potential in the geographical information industry (Greve, 1996).

With the stereo-model reconstituted, and assuming the availability of digital images, it is possible to produce digital ortho-images or ortho-photos, which are supposedly free of image displacement due to photograph tilt and terrain relief. The U.S. Geological Survey lists digital ortho-photo quadrangles (DOQs) as one of its core digital geographical data products. The production of ortho-images is also known as image rectification, and may be done in a direct or indirect mode (Konecny, 1979). The so-called direct method for image rectification applies coordinate transformations from an original image to its rectified version using

Equation 3.17 with knowledge of image orientation. However, the rectified image points are no longer spaced at regular intervals. An indirect method for rectification seems a more natural way to produce DEMs, especially those derived from digital photogrammetry.

Figure 3.9 shows the process of indirect rectification of digital images, in which a known grid node J with coordinates $(X,Y,Z)_J^T$ is transformed using Equation 3.17 backwards to point j in the original image for retrieval of its image grey level. As the transformed image coordinates for point j may not be integers, a process called image resampling is performed to interpolate grey levels within the neighbourhood of digitised image pixels. The nearest neighbour approach may be used to assign the grey level of the pixel closest to the image point, and this approach has very modest computational complexity. Bilinear interpolation may be used to gain improved results at the expense of more complicated computing, while bicubic convolution is even more sophisticated and costly but achieves even better accuracy.

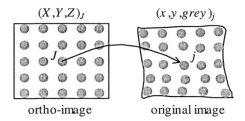

$(X,Y,Z)_J$ $(x,y,grey)_j$

ortho-image   original image

Figure 3.9 Generating an ortho-image.

The accuracy of elevation data derived from photogrammetry relies on the quality of the image matching, as can be seen from Equation 3.18, where the accuracy of x-parallax measurement has a direct bearing on the derivation of height. Similarly, one may be able to analyse the degree of approximation involved in image digitisation, rectification, and resampling for digital ortho-image generation, though sub-pixel accuracy is often attainable in operational settings.

A useful approximate formula for evaluating errors in photogrammetric height data is based on an alternative form of Equation 3.20, such that differentiation in $H_{J\text{-}J0}$ is approximated as:

$$dH_{J-J0} = H_{J0}dpx_j/base \tag{3.30}$$

where *base* is the baseline length $\|S'-S\|$ at photo scale (Hallert, 1960). The effects of errors in x-parallax difference measurement on height derivation may become serious if stereoscopic viewing by humans and image matching by computers are not of sufficient accuracy. While it is often believed that measurement by trained operators is reasonably accurate for a given image scale, computer stereoscopic viewing based on image correlation may be questionable because of low *SNR* images and inappropriately designed algorithms, among other factors. Ehlers (1985) evaluated the probabilistic effects of image noise on digital correlation using simulated noises superimposed on a digitised aerial photograph.

Image matching may be enhanced by utilising a certain amount of image reasoning and artificial intelligence, and interesting developments may lie in this direction. One approach is to extract objects from imaged scenes, and to obtain three-dimensional models of the extracted objects. This is particularly useful in urban modelling, where built structures and cultural landscapes pose challenging problems for digital plotters, which have to distinguish between the genuine terrain and the apparent terrain as modified by structures. When there are no ambiguities regarding the true terrain, DEMs can then be derived without the inaccuracies imposed by artefacts, which can actually be mapped as discrete objects in their own right. Another direct result of object extraction is that digital ortho-images can be generated more accurately than otherwise, since the digits of pixel grey levels are treated intelligently instead of arbitrarily.

When objects and terrain are mapped accurately, it is possible to see and interpret more useful information from images, since these are now rectified and may be used as ortho-projected maps. Thus, it becomes possible to map out geographical distributions of quantities and qualities by utilising image sources with enhanced spatial, spectral, and temporal resolution. For this, analysis of multi-resolution images captured by various remote sensors may prove feasible.

3.4 REMOTE SENSING

A substantial amount of thematic information can be obtained from aerial photographs by applying photo interpretation techniques, in addition to the geometric measurements of object positions, sizes, perimeters, areas, and volumes that form the basis of topographic mapping based on a reconstructed stereo-model. Aerial photo interpretation is a process by which information on site, association, tone, geometry, texture, pattern, size, and shadow is utilised by an interpreter to obtain reliable identification of entities on the ground. Further details of the nature of aerial photo interpretation are beyond the scope of this book.

Though not explicitly documented in the previous section, a variety of aerial photographic techniques exist, providing different ways of deriving geometric and thematic information. Panchromatic film for aerial photography is being progressively superseded by natural colour film, which is widely used in urban applications because of the associated ease of interpretation: the colour balance parallels the interpreter's real-world experience. Colour infrared film is particularly valuable because of the increased contrast between vegetation, water, and man-made structures, but is less suitable for general-purpose interpretation.

A constellation of satellites with sensors aboard has captured a vast volume of digital image data over the Earth. Such sensors are widely acclaimed for their synoptic coverage, greatly increased spectral resolution, and rapid updating capability. Remote sensors of high spectral resolution, in hundreds of spectral bands with widths of less than 5 nm, have been used for species discrimination and vegetation stress detection (Ferns *et al.*, 1984). Repeated global coverages of remote sensor data are making increasing contributions to geographical information infrastructure.

For digital images acquired from scanning systems, spatial resolution is determined by the so-called instantaneous field of view (IFOV), which is approximately 0.086 milliradians for spaceborne scanning systems. For airborne scanning systems, IFOV is generally in the range between 1 and 2 milliradians. Within the IFOV, spectral response does not ideally conform to a point response function exactly but may be approximated by certain kinds of point spread functions (PSFs). Thus, recorded radiance $z(x)$ for a pixel x should be viewed as the convolution of a specific $PSF(x)$ and the true radiance $Z(x)$ plus an additive noise term *noise*(x), as:

$$z(x) = PSF(x)Z(x) + noise(x) \tag{3.31}$$

Carnahan and Zhou (1986) used Fourier transform techniques for quantifying the PSF of the Landsat TM sensor. They considered the function to be separable, leading to an evaluation of line-spread functions in along- and cross-scanning directions. It is important to recognise the degradation of the radiance actually acquired and recorded through radiometric errors, which contaminate remote sensor data and present a challenging problem. However, as is the case with any measuring system, this physical limitation should not discourage applications of remote sensing technologies.

Table 3.1 lists the main features of some common remote sensing systems currently in orbit. Not surprisingly, the ground resolution of satellite images is considerably inferior to that of aerial images, and spatial resolutions of 1 metre and better are easily obtainable in the latter case. However, it must be emphasised that different applications are served well with remote sensor data of varied spatial resolution, and accuracy should be assessed in the spirit of fitness-for-use. A spatial resolution of 0.5 metres would still be inadequate for surveying property boundaries, while satellite images of 1 kilometre resolution could suffice for a global inventory of tropical forestry.

Table 3.1 Optical and radar sensors on spaceborne platforms.

Platform	Sensor	Bands	Ground resolution (m)
Landsat	MSS	4	79
Landsat	TM	7	30 120 (ch. 6)
SPOT	HRV	4	10 (P) 20 (XS)
ERS-1	AMI	*C*-band	30
RADARSAT	SAR	*C*-band	25×28 11×9

As is shown in Table 3.1, two kinds of sensors are relevant for mapping: optical and microwave. The optical sensors, such as Landsat TM and SPOT HRV, detect reflected sunlight and re-emitted thermal radiation. On the other hand, microwave sensors, such as the active microwave instrumentation (AMI) on board ERS-1 and the synthetic aperture radar (SAR) on board JERS-1 and RADARSAT, deal with both transmission and reflection of energy in the microwave portion of the radio-frequency spectrum.

Recent developments in radar technology have been led by the emergence of interferometry. Typically, radar measures the strength and the round-trip time of a radio wave emitted from an antenna and reflected off a distant object. Interferometry radar combines two sources of radar that beam microwaves in two separate passes over the Earth's surface and record the backscattered reflection from points on the ground. Using this technology on a recent space shuttle mission, a research project combining efforts from the U.S. National Aeronautics and Space Administration (NASA) and the National Imagery and Mapping Agency (NIMA) made elevation readings of most of the Earth's surface every 30 meters, taking topographic mapping to a new level of spatial resolution and comprehensive coverage. Discussions of radar technologies may be found in the remote sensing literature (e.g., Curlander and McDonough, 1991), and the performance and limitations of different methods for radar-based elevation extraction have been reviewed by Toutin and Gray (2000). This section, however, concentrates on key issues such as the geo-referencing and classification of remotely sensed data.

Geo-referencing of remotely sensed images may follow the approach of ortho-image generation by photogrammetric rectification, as shown in Figure 3.9. However, unless perspective displacement due to terrain undulation is significant for a chosen scale of mapping, many remotely sensed images can be rectified using polynomials linking the image with the ground coordinates of a few GCPs. For example, given a number n of GCPs, the following equation is commonly used to transform images arrays to ground coordinates:

$$X_J = a_0 + a_1 x_j + a_2 y_j + a_3 x_j y_j + a_4 x_j^2 + a_5 y_j^2$$

$$Y_J = b_0 + b_1 x_j + b_2 y_j + b_3 x_j y_j + b_4 x_j^2 + b_5 y_j^2 \tag{3.32}$$

where coefficients a_0 through a_5 and b_0 through b_5 are to be solved by least squares on the basis of a few well-distributed GCPs with known coordinates (X_J, Y_J), which are also identifiable on the image plane with coordinates (x_j, y_j). Frequently, positional accuracy at sub-pixel levels is reported in empirical studies.

In regard to mapping geographical distributions, both continuous and categorical variables can be estimated from the analysis of remotely sensed data. Dozier (1981) applied remote sensing techniques to identify surface temperature fields from satellite images with sub-pixel resolution. As has been mentioned in Chapter 2, certain spatially distributed variables such as biomass may be directly related to multispectral image data via defined quantities such as NDVI, defined in Equation 2.8.

Quantities estimated through the use of such methods will be inaccurate due to radiometric errors in the relevant bands of spectral data. Error propagation from radiance data to end estimates may be quantified analytically. But enormous difficulties remain for interpreting radiance data as arrays of digital numbers (DNs) in relation to the geographical variables under investigation. This is because radiance data have undergone complex mechanical, optical, and electronic transformations, during which errors may be sufficiently poorly understood to render the remote sensing of spatial distributions liable to great uncertainty (Cracknell, 1998).

Information may also be extracted from remote sensor data concerning the distributions of geographical categories such as land cover and soil type. An increasing amount of categorical data is derived from the computer classification of remotely sensed images, especially those in digital form. There are two main approaches, supervised and unsupervised, to the classification of remotely sensed images, as follows.

Supervised methods use ground truth data, that is, direct observations of conditions on the ground (Steven, 1987). They begin with a compilation of training data for each class to be mapped. A skilled image analyst works on the screen, selecting areas known to contain the classes or categories named. Training pixels provide spectral signatures for all categories in the classification scheme chosen. Spectral means and standard deviations are calculated for each class. An illustration is provided in Figure 3.10, where three categories U_1, U_2, and U_3 are unusually well separated in the spectral space consisting of two bands of digital numbers.

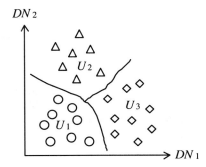

Figure 3.10 Spectral signatures of three geographical categories.

Upon establishment of the spectral signatures for the desired categories, a classifier comprised of certain rules or algorithms is designed so that pixels of unknown classes can be classified with maximum accuracy. Different classifiers may be implemented with different performance. A classifier can be based on the criterion of minimum spectral distance, by which an unknown pixel x is assigned · to its nearest class mean in the spectral space.

Denote the categorisation of pixel x as $u(x)$, and the spectral means of candidate classes as m_k for a total number of c classes $\{U_1, U_2, \ldots, U_c\}$. Then a distance-based classifier decides that a pixel should be assigned to class U_{k0}, that is:

$$u(x) = U_{k0} \tag{3.33}$$

if $distance(DN(x), m_{k0}) < distance(DN(x), m_k)$ for all $k \neq k0$, where vector $DN(x)$ has components DN_1, DN_2, etc.

More probabilistically sophisticated methods such as maximum-likelihood classifiers or Bayesian decision rules can be used in remote-sensing applications when *a priori* knowledge of the likely relative frequency of classes is available. In particular, maximum-likelihood classifiers assume that the training sample of each class has a multi-variate normal probability density distribution in spectral space, defined by the position of the centroid and the variance-covariance matrix of the sample distribution (Townshend, 1981).

Unsupervised classification, as an alternative to supervised classification, exploits the inherent structure in the data. It starts with a cluster training procedure based on spectral characteristics. Clustering is a way of ordering data by sorting pixels into classes according to a distance measure. The analytical procedure for clustering may be interactive, and includes iterative steps of: a) initialising the clusters; b) assigning pixels to the nearest cluster according to a metric of spectral similarity, or distance in spectral space; and c) calculating updated clusters means. The process stops when the clusters no longer change between successive iterations. The characteristics of the clusters are determined *a posteriori* by looking at the ground properties of samples from individual clusters (Swain and Davis, 1978).

The unsupervised method is especially useful when working in a new area, because the clustered classes should reveal what land cover or terrain types can most successfully be distinguished by using the image data. It is strongly recommended for use in regions of natural, non-agricultural terrain, because of the difficulty of obtaining unique spectral signatures or homogeneous training samples. The difficulty may be due to spectral differences in the cover types themselves, or variations in the terrain. However, the resultant classes from an unsupervised method are not guaranteed to be useful. Some of the clusters may be meaningless because they may contain too wide a variety of ground conditions. Moreover, the interpreter will make less use of the ground and ancillary data that are usually available for interpretation, implying that the classified data will have limited accuracy.

In overview, the accuracy of the classification of remote-sensing images, either by supervised or unsupervised methods, is affected by many factors, just as errors creep into the measurement of continuous variables by remote sensing. Both supervised and unsupervised classifiers assume that the image data form separate groups in spectral space, and the groups can be associated with ground observations. These groups are demarcated by boundaries, such as the rectangular shapes assumed by parallelepiped classifiers, and the hyper-ellipsoid shapes assumed in maximum-likelihood classification. If such well-defined spectral signatures could be identified for geographical categories, the digital classification of remote sensor images would be performed with reasonable accuracy and replicability (Richards, 1996). However, according to Skidmore and Turner (1988), it is frequently observed that clusters in spectral space fail to exhibit distinctive patterns, and that such patterns as do exist fail to be approximated by rectangular or hyper-ellipsoid shapes. Therefore, classification uncertainties are bound to exist in remote sensing data.

Complication and inaccuracy of classification frequently occur due to the presence of mixels (mixed pixels, that are not completely occupied by a single, homogeneous category). For example, the heterogeneous nature of urban areas (i.e., the varying mixture of urban surfaces) produces a mixed pixel response, as shown by Forster (1983), even with data of higher spatial resolution (Mather, 1987). Campbell (1996) showed that mixed pixels are common especially at the edge of large parcels, or along long linear features, or among scattered occurrences of small parcels. Besides, mixed pixels also occur where the land cover elements are continuous and gradual, as reviewed by Cracknell (1998).

Though the mixed pixel contains more than one class, it can only be allocated to one class. Furthermore, as the mixed pixel displays a composite spectral response, which may be dissimilar to each of its component classes, the pixels may not to be allocated to any one of the component classes. Error is therefore present in the classified image when mixels are present. Any estimation of the areal extent of the land cover classes may thus also be prone to error. In short, mixed pixels degrade the accuracy of image classification.

A growing body of research into new methods for classification of remotely sensed images based on newer contextual, neural network, and rule-based approaches may overcome some of the shortcomings in supervised and unsupervised approaches and adapt to the uncertainties occurring in remote sensing techniques. However, no method can claim to be perfect (Davis and Simonett, 1991). And, given the fact that remotely sensed images have undergone only approximate corrections for radiometric and geometric distortions, any use of remote sensing data must address error issues.

In summary, the multispectral data of airborne or spaceborne sensors are used for natural-resources mapping and environment monitoring based on image processing systems. However, remotely sensed data continue to suffer radiometric and geometric distortions. Besides, the relatively large pixel size of the sensors implies that there is a high probability of more than one cover type contributing to the recorded reflectance values. Moreover, individual patches of the same category may have different spectral signatures, as Bailey (1988) showed for ecological applications. The net result of these problems of sampling resolution and complex relationships between ground conditions and recorded signatures is that it is difficult to obtain classification accuracy better than 70% for any class on an image other than water, as has been consistently shown by various studies (Stuart, 1996, pers. comm.).

3.5 MAP DIGITISING

Land surveying, photogrammetric, and remote sensing techniques perform measurements and collect data directly from the field, from photographic and digital images. They are considered as primary data acquisition methods, and they lend themselves to current, reliable, and direct measurements (Thapa and Burtch, 1991). On the other hand, depending on the specific requirements of projects, data may be acquired with sufficient accuracy from existing sources, such as maps, charts, or graphs, which are considered as secondary products derived from the primary products.

Map digitising, for converting paper maps into digital databases, is probably one of the most widely used secondary data acquisition methods. For instance, the digital line graph (DLG) data routinely produced by the U.S. Geological Survey (USGS) are digital representations of cartographic information. They contain a full range of attribute codes and are topologically structured after passing certain quality-control checks. DLG data are available in a range of scales. Large-scale DLG data are available for the U.S. Public Land Survey, boundaries, transportation, hydrography, and in some areas, hypsography, and are produced in 7.5-minute by 7.5-minute tiles corresponding to USGS 1:24,000 scale topographic quadrangle maps. Intermediate-scale DLG data (1:100,000 scale) cover transportation and hydrography, and are available in 30-minute by 30-minute tiles. National Atlas tiles of the U.S. at 1:2 million scale are available for boundaries, transportation, and hydrography.

The most commonly used map digitising method is manual digitisation, performed by locating a cursor at the point or moving a cursor along the line to be digitised. The method also requires software to display and store the digitised data. In general, digitising is performed on a stable-based copy of the graphic material, which is mounted on a digitising table with a fine resolution of, say, 0.001 inches. Four control points, known as tics for ARC/INFO and corresponding to the four corners of the quadrangle, are used for registration during digitising so that appropriate transformations can be performed from the internal coordinates of the digitiser to Universal Transverse Mercator (UTM) grid coordinates. Attribute data entry is performed either at the same time as map digitising, or on an interactive edit station after completion of digitising. A digitised map is often plotted out for checking against the original map being digitised.

After decades of GIS database creation, the technical processing involved in map digitising has become common and widely accessible. However, techniques for improving digitising efficiency and accuracy are always welcome. As might be expected, densely urbanised areas are mapped at greater scales than the sparsely populated areas. Many built structures possess certain geometric regularities, which may be exploited to boost the accuracy of digitised data.

Figure 3.11 shows a hypothetical example featuring a segment of highway and a six-edged house. As can be seen from Figure 3.11, the regularities of the house's outline consist of pairs of parallels and perpendiculars: edges x_1 to x_2, x_3 to x_4, and x_5 to x_6 are parallels, and so are edges x_1 to x_6, x_2 to x_3, and x_4 to x_5; all adjacent edges intersect at right angles.

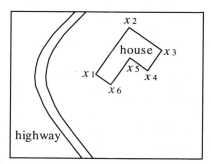

Figure 3.11 Digitising map objects of regular shape.

Suppose all six vertices defining the house shown in Figure 3.11 are digitised. There will be five redundancies out of twelve measurements, based on two coordinates for each vertex. These five redundancies can form five conditions, either in the form of five rectangles or as a combination of four parallels plus one rectangle. For illustrative purpose, the rectangle formed by edges x_1 to x_2 and x_2 to x_3 is expressed as:

$$(X_1 - X_2)(X_3 - X_2) + (Y_1 - Y_2)(Y_3 - Y_2) = 0 \qquad (3.34)$$

where the coordinate pairs for vertices x_1 to x_3 are annotated respectively with subscripts. On the other hand, the parallel pair, edges x_1 to x_2 and x_3 to x_4, forms the following condition:

$$(X_1 - X_2)(X_3 - X_4) - (Y_1 - Y_2)(Y_3 - Y_4) = 0 \qquad (3.35)$$

based on the coordinate pairs of vertices x_1 to x_4.

Least squares on the basis of the above conditions can provide most probable estimates for all the six vertices marking the house footprint shown in Figure 3.11. Similarly, raw digitising data regarding the highway in Figure 3.11 may be adjusted by exploiting the implied conditions of parallelism or equal width. In manual stereoscopic digitising of aerial photographs, the same technique may be usefully applied. In addition, when image segments are hidden or overshadowed by features such as full-grown trees, certain objects of regular shape may still be recovered accurately by using geometric constraints such as those above.

Digitising methods may be selected based on various criteria, including map feature density and the availability of data production technologies. Automated line-following techniques are devised to automate manual digitising to some extent. They are widely used to digitise linear features, for example, rivers, roads, and contour lines (Sircar and Cebrian, 1991). To minimise inaccuracy in the line-following process, different map separates may be used. Map or image scanning involves a computer-controlled instrument equipped with optics and detectors that can create the digital data from the document. There are several types of scanners, including drum scanners, laser beam scanners, and video-camera scanners. The scanning of a document may take only a few seconds, but the subsequent vectorisation and editing will take longer to complete.

The process for generating a USGS digital raster graphic (DRG) is as follows. A digital image is produced by scanning a USGS map on a high-resolution scanner at a resolution of 500 dots per inch (dpi), resulting in a red-green-blue (RGB) file of up to 300 megabytes. The RGB file is then viewed, and the position of each of the sixteen 2.5-minute grid ticks on the image is collected, and their coordinates are used for rectifying and georeferencing the image to the UTM ground coordinates through piecewise linear rubber sheeting. The image is resampled to 250 dpi and compressed to reduce the size of the data set, resulting in a data file typically sized between 5 and 15 megabytes.

Digitised data from maps will suffer any uncertainties that are already embedded in the source maps themselves, and errors introduced by the digitising process (Bolstad *et al.*, 1990). The former source of errors exists due to the fact that published maps have usually been subjected to a map generalisation process, which implies a variety of abstraction, selection, simplification, and approximation steps (Brassel and Weibel, 1988; João, 1998). Thus maps portray selected features of reality in a highly abstracted and generalised form, a point made by Harley (1975). Some maps can be comparatively free of impurity, such as property parcel maps. In geographical data on natural resources such as vegetation, soil, or land use, a major cause of errors is the omission of heterogeneous mapping units within areas delineated on the map as uniform (Chrisman, 1989). This problem is caused by the oversimplification of a more complex real world.

As for other map objects, the density of contours representing terrain elevation is directly related to the complexity of the underlying terrain surface. This implies that hilly areas exhibit more variability than flatter areas, and will require finer resolution in data capture than the latter. However, contours digitised from paper maps may be treated differently from real linear and polygonal features, which should be digitised with a level of positional accuracy that is possible at the scale of source maps and the resolution of processing. As a type of field model, contours should be digitised and processed with the primary concern over accuracy in elevation. This means that certainty in elevation representation may not be undermined that much by adaptive selection of points along contours rather than taking their whole, whereby more points are captured from areas showing increased variation than those with relative homogeneity. Nonetheless, at a higher level, morphological characteristics of contours may be usefully incorporated for surface representation matching the effects of generalisation. Map objects that are abstracted from natural variations may be treated similarly, distancing them from real world features that demand accurate positioning and attribution in GIS.

The digitising of maps, either by manual or by scanning methods, is subject to a host of inaccuracies imposed by machine and operator limitations, though scanning methods may be more consistent and controllable in regard to positional errors than manual methods. Dunn *et al.* (1990) discussed the uncertainties that were introduced during their digitising of land-use data from maps. Uncertainties created during map generalisation are not easily identified and measured, unless more information is provided with respect to the specifications and procedures applied. Therefore, the distinction between primary and secondary data acquisition is useful, because uncertainties in the former are easier to identify and quantify than in the latter. Furthermore, maps may have varying levels of mapped detail, resulting in the slivers encountered in map overlay processing (Goodchild, 1978; Filin and Doytsher, 1999).

In summary, although map digitising remains a viable and reasonably accurate method for GIS data collection, there may be too wide a gap between what is captured on a map and what exists in reality. Thus, despite the careful use of large-scale map data for registering collateral data sets or rectifying remote-sensor images, map digitising is generally no substitute for primary data collection, especially when geographical features and distributions are non-homogeneous and poorly defined. Despite a long history of efforts to make map-making accurate, the paper maps that form the source for GIS databases inevitably possess substantial uncertainties, and these are compounded by additional errors introduced during the digitising process.

3.6 DISCUSSION

This chapter has shown that various technologies are operationally available for cost-effective production of geographical data. Such data are crucial in many applications. For example, three-dimensional digital maps are used for military applications such as improved flight simulators, logistical planning, missile and weapons guidance systems, and battlefield management and tactics. Digital maps available to the public, environmentalists, urban planners, and administrators provide the basis for the management of roads, buildings, forests, and wetlands, and the allocation of resources.

However, the production and use of geographical data has to face and cope with substantial uncertainty and opportunity for error. Whether by fixing positions using a total station, employing GPS-based positioning, or interpreting remotely sensed images, the geographical world is sampled and measured so that its most relevant aspects are effectively mapped prior to spatial analysis and decision-making. For instance, a set of samples, such as soil pH values, is designed to characterise the significant variations that are present in a larger universe, that is, at all possible locations within the mapped domain. On the other hand, discrete objects, such as roads, may be identified, positioned, and attributed on a photogrammetric workstation.

Thus, for complete understanding of inaccuracy and, hence, fuller use of uncertain geographical data in practice, geographical abstraction and measurement are both critical dimensions for the analysis of the uncertainties present in any data sets undergoing geo-processing. A lack of recognition of either component often leads to misinterpretation, unrealistic expectations about uncertainty, and, in general, a failure to understand the consequences of using data that are uncertain to varying degrees.

Geographical modelling has been treated in Chapter 2, and measurement has been the subject of this chapter. Heuvelink (1998) identified the subtly distinctive aspects and complex interactions of these two processes. Uncertainty resulting from geographical conceptualisation is reflected in variability, and the structures of errors modelled by covariance are too important to be omitted in any statistical analysis of geographical distributions.

Geographical conceptualisation influences the ways sampling may be performed. Within a geographical domain, there are several standard approaches to sampling. In a random sample, every place or time is equally likely to be chosen. Systematic samples, often acquired by remote sensors, are chosen according to a rule, for example, every 1 km or on a daily basis, and the rule is expected to create no bias in the results of analysis, or no significant difference from a truly random sampling. In a stratified sample, the universe is decomposed into significantly different sub-populations, within which samples are collected in order to achieve an adequate representation of each. For example, it may be known that the topography is more rugged in one part of the area than in other parts, and samples are taken more densely there to ensure adequate representation. The implication of sampling is that geographical complexity is selectively measured rather than copied precisely into digital form. This is something that users of geographical data must always bear in mind.

Errors also result from the use of a variety of information technologies. For example, despite the widespread application of photogrammetric techniques for the provision of positional and thematic information, various uncertainties occur in all stages of photogrammetric data acquisition and processing. These uncertainties occur from aerial photography to the final data output, due to camera lens distortion, atmospheric refraction, displacement due to tilt and relief, instrument limitation, and operator bias. The accuracy of digitising from a correctly oriented stereo-model will depend not only on the stereoscopic acuity of the operator at work, but also on his or her interpretation. Often, interpretation bias has greater effects on digitising results than the accuracy of measurement, as Kirby (1988) discussed in the context of aerial survey of a Scottish rural landscape. Rosenberg (1971) long ago challenged the traditional view that photographic quality will satisfy most requirements for mapping, since the value of photogrammetry extends well beyond point-based measurement to object and scene description, that in turn requires detectability and recognisability. Error issues seem even more acute in digital photogrammetric systems, where data production is highly automated. Unverified image matching may lead to errors in the resultant three-dimensional data and image maps, demonstrating that computing is far from a complete replacement for the human stereoscopic analysts (Torlegård, 1986).

Uncertainties also surrounded remote sensing, both in image data acquisition and in subsequent data handling. Radiometric and spatial distortions are present, and the measured radiometric value of each pixel may be degraded by noise caused by sensor inaccuracy, and variations in reflected energy due to changes in terrain and solar incidence. Geometric distortions are created by the rotation of the Earth, by instability of the satellite in orbit or the aircraft in flight, by atmospheric aberrations, and by variations in surface elevation, as described in detail by Schowengerdt (1997). Errors are inevitable, whether remotely sensed data are used for variable estimation or for spatial classification, although useful methods continue to be devised to reduce their influence.

Errors in different data products may actually be closely correlated. For instance, errors are likely to compound if DEMs and DOQs are produced not in isolation but in sequence, especially when both are generated from photogrammetric sources (Wrobel, 1991). Thus, error behaviours of DOQs are directly affected by how accurately DEMs are created from image pairs and how effectively they are used to correct for image displacement, if the latter is usually used to support creation of the former. Similar examples may be found in land classification, whereby interpretation of aerial photographs is performed to support and validate semi-automatic classification of satellite images. As photo-interpretation is not perfect, performance and accuracy assessment of satellite image-based classification, which relies on training and testing data generated from aerial sources, will be open to question, even if image classifiers are verified to be reasonably accurate in laboratory environments.

Finally, any method for data reduction is approximate, just as measurement is always less than perfect, capturing only part of the real-world variation. Positioning discrete objects is not immune to errors either. No matter whether objects are closely associated with real-world entities or exist only as conceptualisations abstracted from reality, there will always be significant limits to accuracy. The combination of heterogeneous data in pursuit of the solution to a spatial problem is complicated by differences in scales, formats, and accuracy. These topics will be explored in later chapters.

CHAPTER 4

Exploring Geographical Uncertainty

4.1 INTRODUCTION

Geographical computing is based on the digital representation, in spatial databases, of geographical features and distributions. Many of the features shown on maps, for example lakes, roads, and human settlements, are conceptualised as discrete objects and are readily represented as such in spatial databases. Land surveyors and photogrammetrists are particularly interested in such well-defined features, which are readily mapped and added to spatial databases as discrete objects with associated geometry, and with attributes derived from qualitative or quantitative descriptions. Such objects are important not only on their own merits, but also for the functions they can perform in co-registering data layers. It has been shown in earlier chapters that the positional and attribute components of spatial objects can be discussed separately, thus explaining the emphasis that is often placed on positional data for discrete objects. Many geometric and topological operations can be performed on objects, by analysing points, lines, and areas with respect to neighbourhood, intersection, and containment.

On the other hand, many spatially distributed phenomena are more effectively conceptualised as fields of continuous or categorical properties. Fields may be used to model physical properties, such as biomass density and vegetational type, that are quantities or qualities defined over space, and may also be used to model variables, such as population density, that are abstracted from social or cultural phenomena. This latter case is exemplified by demographic mapping and its dependence on ground-based survey of socio-economic data for the analysis and prediction of processes related to population. In the domain of continuous and categorical fields, spatial variability and dependence are the two interrelated characteristics that are of particular importance because of their potential for insight. Quantification of these two aspects is often at the core of field-based geographical analysis, since many spatially distributed phenomena exhibit both pronounced heterogeneity and pronounced dependence.

Various kinds of inaccuracy or, in a broader sense, uncertainty have been discussed in previous chapters as being endemic to geographical information, and as affecting the applications that depend on geographical information. Specifically, the process of geographical abstraction or generalisation about real world phenomena imposes degrees of approximation and discretisation that are necessary for the digital implementation of spatial problem-solving, while geographical measurement suffers from errors in sampling and mapping. Gahegan and Ehlers (2000) discussed the accumulation of uncertainty through the transformation of raw remote-sensor images to GIS data.

The ubiquitous nature of uncertainty suggests that comprehensive taxonomic description will be required in order to understand the mechanisms of uncertainty occurrence and propagation. Some well-established statistical methods and tools for error analysis may be of use in handling geographical uncertainty, although how effective they can be is questionable. Gaussian models seem to provide a suitable starting point for describing and modelling quantitative data sets, given the law of large numbers and the central limit theorem (Kolmogorov, 1956).

Point coordinates are traditionally derived from the precise measurement of distances and angles, and their adjustment using least squares if there is redundancy in measurement, as has been shown in Chapter 3, which covered a variety of techniques for determining position. The covariance of coordinates derived through this process may be quantified through variance propagation, which may be further used for assessing accuracy in any data sets constructed from the coordinates (Drummond, 1995).

Accuracy can also be assessed by checking samples of points against ground truth, and constructing appropriate error measures of positional accuracy, that can then be generalised to the entire database. Epsilon error bands are often considered as a simple extension of point positional errors to lines and areas. Further, as areas are formed as polygons from lines or, in raster, from contiguous groups of grid cells, areal inaccuracy is likely to be reflected in boundary uncertainty (Crapper, 1984).

However, geographical information is much richer than is implied by a database of discrete points, lines, and areas. Fields are the other alternative for modelling those geographical distributions that are effectively captured in the form of continuous or categorical variables. Just as for positional data recorded as coordinates on continuous scales, variables measured on continuous scales, such as surface temperature and terrain slope, are obviously also amenable to statistical error analysis. Metrics of root-mean-square error (RMSE) are useful indices of errors in continuous variables.

Categorical variables, on the other hand, take discrete values as the outcomes of classification processes. The results of accuracy assessment are likely to be in the form of agreements or disagreements between classified and reference categories. Classification accuracy may be evaluated by measures such as the percentage of correctly classified objects (PCC) and the kappa coefficient of agreement, a topic reviewed by Congalton (1991), Goodchild (1994), and Congalton and Green (1999). Accuracy measures such as PCC can be usefully related by end users to actual categories through probabilistic interpretations.

The rigor of mathematical and statistical treatment of prominent point objects such as GCPs and photo-control points is by no means conclusive, given the complexity of the geographical world. Common sense suggests that digitised objects will be no more accurate in positioning than the reference points that have been used for geo-referencing and model orientation. In many circumstances it may be necessary to recognise that accuracy can vary spatially, such that higher accuracies are obtainable in some areas than in others.

Similar observations are applicable for fields of both continuous and categorical variables. With an accuracy-oriented paradigm, accuracy measures such as RMSE and PCC are actually non-spatial metrics adopted for spatially distributed errors. It is reasonable that locations closer to samples should have less uncertainty than those that are located further away. This model of spatially varying uncertainty is clearly quantified by Kriging variances, the by-products of using Kriging for spatial interpolation, as described in Chapter 2.

In classifying remotely sensed images, ground truth data are often required for training a supervised classifier. They are usually selected from well-controlled and known locations, where categories are believed to be the purest and most representative. As the contents of training pixels will be used to separate the classes, it is possible assert that a classified image will exhibit varying levels of uncertainty over space. Again, the general patterns of uncertainty will be spatially dependent, with locations closer to training pixels having smaller inaccuracy in classification than those further away. Similar remarks were made by Thomas and Allcock (1984), who also emphasised the need for the quantification of confidence levels in classified data.

The recognition of spatial variability of errors in various types of geographical data leads to further thoughts about uncertainty modelling, which represents a major technique for better geographical information. Through such modelling it is possible to assess the consequences of using data of varying accuracy before they are put to actual use. Traditional statistical approaches and their simple extensions are hardly adequate for error modelling, which is substantially more complex than error description, and requires knowledge of spatially varied and correlated distributions of errors. For example, even with knowledge of spatially varying accuracy for points, lines, and areas, it is difficult to determine confidence intervals for lines and areas without taking error correlation into account. This is particularly relevant to map digitising, which is known to produce serial or spatial correlation in the errors in digitised points, such that the error at one point in a digitised sequence is strongly related to errors at adjacent points in the sequence. Spatial correlation in positional data should thus be properly accounted for in uncertainty handling.

For modelling rather than merely describing errors in terms of statistical and probabilistic indices, it is not sufficient to deal with measurements of spatial location in isolation; rather, one has to quantify spatial variability and correlation so that geo-processing and derivative information will be evaluated in a way that is objective and correct. Field-based uncertainty analysis can make use of developments of geostatistics, since this area includes a selection of techniques for modelling the spatial variability and dependence that are characteristic of geographical fields, no matter whether the field variables are continuous or categorical.

The foregoing arguments have been placed in a context of measurements and observations that are implicitly assumed to be of a crisp nature. From a vector database, it is possible, for example, to calculate the precise distance between a potential site for a waste dump and a recreational park. When classification of land cover is conceptualised as a categorical field, it is assumed that individual locations can be assigned on a yes or no basis. However, in reality much vagueness is involved in such examples. Geographical distances are often known only approximately, between objects whose definition and existence may be unclear. Similarly, land classification may be a vague and uncertain process of deciding between poorly understood categories. Although geographical databases have the appearance of precision, the human context in which they are created is often extremely vague.

Fuzzy set theory has been created to imitate, represent, and reason with the vagueness inherent in much human cognition and intelligence, and has encountered strong interest in geographical analysis and applications (Zadeh, 1965; Pipkin, 1978). In brief, fuzzy sets can be seen as a body of theory and methods that is complementary to classic probability theory, in the sense that it seeks to describe and interpret the uncertainties that result from vagueness rather than randomness, as the following sections explain.

4.2 GEOGRAPHICAL UNCERTAINTY DESCRIBED

The processing and handling of digital geographical information in computers have prompted a prolonged debate over efficiency and accuracy (Berry, 1987). Often, computing permits a level of numerical precision that is much higher than the inherent accuracy with which geographical phenomena are described and geographical data are acquired. This apparent gap between data inaccuracy and computing capability has attracted a growing level of research interest that is aimed at achieving a better fundamental understanding of geographical uncertainty, and of the processes of production and use of geographical data (Goodchild and Gopal, 1989).

Fortunately, probabilistic theories and statistical methods are already well established, though primarily for use in non-spatial contexts. Such techniques are available to geographical information scientists and analysts for application, where appropriate, to the modelling of errors in geographical data (Zarkovich, 1966; Gregory, 1978; Rosenfield and Melley, 1980). This section begins with a brief review of the concepts of probability and statistics of relevance to the geographical context.

4.2.1 Background to Error Analysis

The concept of population is an important one in statistical inference, which is defined as the totality of observations of interest. The number of observations in the population is defined to be the size of the population, which can be finite or infinite. The population count of a city is a finite number, while measurements of atmospheric pressure anywhere on the Earth's surface, or the depth of a lake from any conceivable position, are examples of populations with infinite sizes. In both cases, and particularly the latter, it is often the case that subsets of observations are relied on to make inferences concerning the population, because it is impossible or impractical to observe the entire population. Even for populations of theoretically finite sizes, subsets are often used as the basis for statistical inference. In either case, samples of various sizes result. If inferences from samples are to be valid, they must be representative of the underlying population. To prevent any possibility of bias in sampling, it is important to choose a random sample that contains observations made independently and at random.

Many types of measurements are effectively governed by randomness. Thus, data, which record outcomes of observations or measurements in statistical experiments either as descriptive representations or as numerical values, are modelled as random variables. For categorical types of random variables, which are measured on discrete scales, uncertainty occurs because realisations of categories within a prescribed classification scheme are not known beforehand. The outcome of a deterministic event is always certain, but probabilistic models allow decision-makers to deal with events whose outcomes are uncertain.

The notation of probability provides the language for reasoning and communication under uncertainty. The mathematical literature on probability is well developed, along with its fundamental principles, axioms, and theorems (Fine, 1973). An event $A \in \sigma$-field is defined as a subset of the sample space Ω, which is a collection of all possibilities or sample points in a probabilistic problem domain. A probability P over a measurable space (Ω, σ) is a real function assigning a measure to an event $A \in \sigma$, such that $P(A)$ is non-negative, and the sum of probabilities over the whole sample space Ω, $P(\Omega)$ is normed to 1. Thus, an event A in terms of probability theory is always referred to as a subset of one or more sample points of a particular sample space Ω.

To quantify the accuracy of spatial categorisation, it is constructive to consider agreements or disagreements between observed and reference descriptions as random events. Knowing the relative frequency of an event, such as correct or incorrect categorisation, amounts to estimating and reasoning with probability. In classical terms, probability is often taken to mean the ratio of the number of outcomes which produce a given event, to the total number of possible outcomes, in an exhaustive set of equally likely outcomes. For a discrete random variable, the probability function gives the probability that a specified event will occur in a certain context.

It is possible to use binomial distributions to describe the occurrence of a category i, if the category's probability of occurrence in any single experiment is given as p_i. To obtain the probability of observing U_i, k times out of a total number of n observations, the binomial distribution is written as:

$$binomial(U_i, k, n) = \frac{n!}{k!(n-k)!} p_i^k (1 - p_i)^{n-k} \tag{4.1}$$

The binomial distribution's expectation and variance are quantified as follows:

$$E(U_i | n) = np_i \tag{4.2}$$

and:

$$var(U_i | n) = np_i (1 - p_i) \tag{4.3}$$

When more than a single category is concerned, as is often the case, multinomial distributions can be specified in an extended form of Equation 4.1 for a number c of space-exhaustive and mutually exclusive categories. Per-category probability must be known somehow in order to use multinomial models. An example is provided in Figure 4.1 where five events (categories) are shown with probabilities $\{0.10, 0.25, 0.30, 0.15, 0.20\}$, which sum to 1.0 to exhaust the event space. Multinomial problems are often decomposed to binomial problems for statistical analysis.

For data derived from measurements and expressed on continuous scales, error (ε) is generally defined as the deviation between the measured or computed data value (g) and the true value or the value accepted as being true (g_0):

$$\varepsilon = g_0 - g \tag{4.4}$$

which may, without loss of clarity, be re-written as the deviation of a measured value from the reference by placing negative signs on both sides.

Continuous random variables or their errors, if given truth values, are characterised by probability density functions, whose integral over a certain interval, that is, the area under the function curve bounded by the interval, gives the probability that a value of the random variable will fall within the interval.

When the intervals expand to infinity, a probability of 1.0 results: in other words, the value of the random variable is certain to lie somewhere.

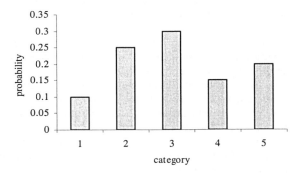

Figure 4.1 A probability distribution function for five discrete categories.

Gaussian models are commonly used with statistical data, especially in large samples which are believed to approximate the normal distribution. A normal distribution for a random variable g is specified with the following probability density function:

$$normal\left(g, m_g, \sigma_g^2\right) = \frac{1}{\sqrt{2\pi}\,\sigma_g} \exp\left(-\left(g - m_g\right)^2 / \sigma_g^2\right) \tag{4.5}$$

where m_g and σ_g^2 represent the mean and variance of variable g, respectively.

Figure 4.2 illustrates a normal or Gaussian distribution of random errors in terms of standard error, that is, $normal(g,0,1)$. In the case of systematic bias the mean error will not be zero, and the curve is shifted accordingly. Based on a specific probability density function, it will be possible to evaluate the probability that errors fall within a certain interval. A well-known value, the probability that normally distributed errors fall within an interval of one standard error either side of the mean, is about 68.3%; approximately 95% of errors fall within two standard errors of the mean.

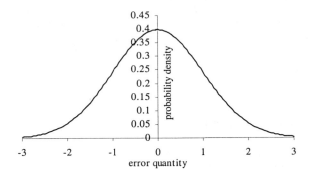

Figure 4.2 A Gaussian probability density function for error quantity in terms of standard error.

It is useful to clarify the concepts of accuracy and precision. Precision measures the degree of conformity of measurements among themselves. For example, the average difference between measurements and their true value would provide a measure of accuracy, while the probabilistic frequency distribution could be used to establish precision, as explained by Bolstad *et al.* (1990). Thus, high precision may not guarantee good accuracy, if measurements are biased. Moreover, confusion and misunderstanding about the distinction between precision and accuracy may become problematic, as spurious precision is at the core of the 'slivers' problem (Goodchild, 1991).

The evaluation of inaccuracy based on Equation 4.4 requires reference or truth data, which are usually taken from independent sources of higher accuracy. This is the most reliable yet demanding method for accuracy assessment. There is actually a hierarchy of methods for estimating accuracy. Deductive estimates may be made in light of the appropriate lineage information. For example, accuracy of a digital coverage created from digitising a 1:2,000 scale topographic map will not be better than 1 m at ground resolution, an estimate implied by the scale of the source map and common standards for map accuracy (a standard of 0.5 mm is a useful rule of thumb for map accuracy, and converts to 1 m on the ground if the map scale is 1:2,000). At a higher level, repeated measurement can provide evidence of accuracy, by taking advantage of the redundancy designed into processes for data acquisition. As for data reduction based on least squares, constraints such as closure of traverse and the expectation that contours will meet at sheet edges may be exploited to provide most probable solutions for map data.

By the law of variance propagation, Equation 3.30 may be extended with a simple example that illustrates how standard error in height measurement can be estimated from known standard errors in x-parallax measurement given specific configurations of base line, camera constant, and flight height. Suppose that an aerial photographic pair is flown with a camera of focal length of 0.10 m at a height of 2,000 m, and that the distance in the air between exposures is 1,000 m. It is possible to use Equation 3.30 to quantify the standard error of height measurement at about 1.2 m, if image coordinates are measured with a standard error of 0.02 mm, a level of accuracy that is reasonably attainable.

In practice, the analysis of errors in geographical scenarios does not usually end there, except for simple-minded problem-solving. This is not hard to see as one recognises the importance of the spatial variability and dependence that are intrinsic to many geographical distributions and that play a crucial role in spatially aware analysis and decision-making. Some approaches to accuracy assessment for spatial data are reviewed below, first for discrete objects and then for fields; accuracies are described descriptively or numerically in statistical and probabilistic terms.

4.2.2 Describing Errors in Objects

There are five fundamental components to the accuracy of geographical phenomena modelled as discrete objects: lineage, positional accuracy, attribute accuracy, logical consistency, and completeness (NCDCDS, 1988). Lineage refers to a description of the source material from which the data were derived, and the methods of derivation involved in producing the final digital data, including details such as the specific control points used and the computational steps taken.

Maintenance of logical consistency may be tested with respect to topology such that, for example, lines connect to each other at nodes and left and right areas are defined consistently for each line segment. Completeness refers to the degree to which the data include all of the relevant and current information. Lineage, logical consistency, and completeness have been discussed in detail by Guptill and Morrison (1995) and are less amenable to analysis than the other two, and thus are not dealt with further here.

Attributes, as qualitative and quantitative facts concerning underlying objects, are often separable from descriptions of the geometric forms and positions of object. Attributes such as tax liability and tenure records for a land parcel clearly function in a way that is independent of parcel boundaries. Moreover, a property may be leased to different tenants, but its position and address remain fixed, although both attribute and position are essential parts of a complete object description in a spatial database. Therefore, as a direct result of the separability of object attributes and positions, attention is concentrated on positional accuracies and how they are described, depending on the types of objects under study.

As specified by the American Society for Photogrammetry and Remote Sensing (ASPRS, 1989), an empirical, site-specific estimate of positional uncertainties can be produced via tests against an independent source of much higher accuracy (this somewhat circular approach is necessary because no reference source can have perfect accuracy). Depending on the specific requirements, the independent source of higher accuracy may be obtained through land surveying or derived from aerial photography. It is specified that the nominal positional accuracy of the reference source or check survey be at least three times that required of the product to be tested. In all tests, it is important to use an adequate number of test data points, sampled in a well-distributed way.

After the check survey, a set of statistical measures of positional accuracy are calculated. Suppose discrepancies or errors ε_i ($i = 1,2,\ldots,n$), observed as the differences in coordinates between the data sets to be tested and the check survey, are quantified using Equation 4.1 for a set of n test points. The sample statistics used for assessing accuracy of map data, applied to the X and Y coordinates, are:

$$m_\varepsilon = \frac{1}{n}\sum_{i=1}^{n}\varepsilon_i \qquad (4.6)$$

and:

$$RMSE_\varepsilon = \left(\frac{1}{n}\sum_{i=1}^{n}\varepsilon_i^2\right)^{1/2} \qquad (4.7)$$

and:

$$std_\varepsilon = \left(\frac{1}{n}\sum_{i=1}^{n}(\varepsilon_i - m_\varepsilon)^2\right)^{1/2}, \qquad (4.8)$$

where m_ε and $RMSE_\varepsilon$ stand for the sample mean error and RMSE, and std_ε is the sample standard deviation of errors.

The sample mean error will be 0 if the discrepancies are unbiassed, but if the sample mean error is non-zero, the sample standard deviation removes the systematic error from each individual error to provide a measure of variation among errors, or precision. Harley (1975) described how the accuracies of maps produced by the Ordnance Survey of Great Britain are tested using this set of equations.

Knowledge of error quantities alone is of limited informational use. The notation of data quality suggests that information on errors should be strengthened with probabilities for statistical inference and decision-making, as is the case with many other applications of quality verification and significance testing. Thus, for one data set that is evaluated with a certain level of error, it is important for data producers and users to be able to compare the level against an error threshold, so that statements regarding data quality may be made based on probabilistic arguments. For example, it is often required that 95% of the sample points must have positional errors less than a prescribed level of 0.5 mm at map scale.

The model of the positional inaccuracy of a point can be represented in the form of an ellipse centred on the point, if the uncertainties in X and Y coordinates are represented by Gaussian distributions, as Goodchild (1991) explained. A Gaussian probability density function, $\text{pdf}(\varepsilon(x))$, may be postulated for positional error $\varepsilon(x)$ at point x as:

$$\text{pdf}\left(\varepsilon(x)\right) = \frac{1}{2\pi\sqrt{\left(\sigma_X^2 \sigma_Y^2 - \sigma_{XY}^2\right)}} \exp\left(-\frac{\varepsilon_X(x)^2 - \varepsilon_X(x)\varepsilon_Y(x) + \varepsilon_Y(x)^2}{2\left(\sigma_X^2 \sigma_Y^2 - \sigma_{XY}^2\right)}\right) \quad (4.9)$$

where $\varepsilon_X(x)$ and $\varepsilon_Y(x)$ represent projected components of the error vector $\varepsilon(x)$ at point x in the X and Y coordinate directions, valued as the difference between an unknown point $x(X,Y)$ and its reference position $x_0(X_0,Y_0)$, that is:

$$\left(\varepsilon_X(x)\varepsilon_Y(x)\right)^{\mathrm{T}} = x - x_0 = \left(X - X_0, Y - Y_0\right)^{\mathrm{T}} \quad (4.10)$$

Clearly, the exponent in Equation 4.9 can be visualised in terms of the elliptical contours, which are possibly slanted if errors are correlated, and σ_{XY} is non-zero. Figure 4.3 provides an illustration for error ellipses centred at a true position x_0, where a two-dimensional probability density function $\text{pdf}(\varepsilon(x))$ is suggested by dashed contours. Using this model, it is possible to compute the probability that the true position lies within any given distance *err* of the measured position, or, the measured position deviates by *err* from the reference position:

$$p\left(\|\varepsilon(x)\| < err\right) = \int\limits_{\|\varepsilon(x)\| < err} \text{pdf}\left(\varepsilon(x)\right) d\varepsilon(x) \quad (4.11)$$

which is evaluated as the integral of $\text{pdf}(\varepsilon(x))$ over the circle shown in Figure 4.3, a volumetric quantity. Similar integrals may be quantified for the probability that the measured position x falls into any specific domain, such as the rectangle shown in Figure 4.3.

When error distributions in both coordinates are the same and independent, the ellipse becomes a circle, again centred on the point. This is the circular normal model of positional error (ASPRS, 1989). When the standard deviations in x and y are not equal, the circular standard deviation is no longer the same as the standard deviations in the two coordinates, but is approximated as the mean of the two. The probability that a point's true position lies somewhere within the circle of radius equal to the circular standard deviation is 39.35%. Under the circular normal model, a radius of 2.146 times the circular standard deviation will contain 90% of the distribution (ASPRS, 1989; Goodchild, 1991).

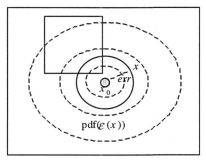

Figure 4.3 Evaluation of probabilities based on probability density.

In vector databases, lines are represented as polylines, or sequences of digitised points connected by straight segments. As has been shown in Chapter 2, spatial objects are often highly abstracted representations of inherently complex entities, implying that errors can be found in any such objects, whether they are discrete points, lines, or areas. The point-in-polygon operation was found to be problematic because of positional inaccuracy in the digitising of administrative boundaries in the British Isles, especially the coastlines (Blakemore, 1984). Positional inaccuracy caused the simple topologic operation to give apparently erroneous results, in which some of the points that should have lain within administrative units were found to lie in the sea.

The epsilon error band model, suggested by Perkal (1956), has been used to describe the inaccuracy of positioning of a digitised line, such that the true position of the line will occur within some displacement ε of the measured position. Thus, the error band width is defined as twice the epsilon distance, that is, 2ε, as shown by the grey-shaded zone in Figure 4.4.

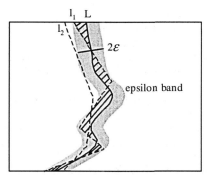

Figure 4.4 Measuring the positional uncertainty of lines.

Epsilon error bands can be used to provide specific tolerance for line positions in GIS operations, such as overlay, and the value of ε can be allowed to vary from object to object. This adjusted form of overlay is more flexible and reliable than the use of a global tolerance specified at the outset, because it can be adapted to the underlying uncertainty levels of lines. Using such techniques it is also possible to generalise the point-in-polygon operation (Blakemore, 1984) to provide three possible results: definitely in, definitely out, and uncertain.

Epsilon error band models have been used in both deterministic and probabilistic forms. In the deterministic form, it is assumed that the true line lies within the band and never deviates outside it. In the probabilistic form, on the other hand, the band is understood as one standard deviation in width, so that one might assume that a randomly chosen point on the observed line had a probability of 68% of lying within the band.

To be used as measures of line uncertainties, the widths of error bands need to be quantified. A sampling scheme, originally proposed for field estimation of forest stand boundaries based on line transects (de Vries, 1986), may prove useful for defining error bands. It is orientated towards applications in areas such as forestry, where reference information is available only in the field and where the cost of survey is relatively high. Thus, it is practically sensible to collect only a sample of the reference source in the form of line transects, from which the separation between the tested and reference positions is determined at each intersection between a transect and the tested boundaries (Skidmore and Turner, 1992; Abeyta and Franklin, 1998). The number of line transects n_1 required for assessing boundary accuracy at a 95% confidence level is related to the total number of intersections of line transects with polygon boundaries n_2 by:

$$n_1 = student^2 / (n_2 error^2)$$
(4.12)

where *student* stands for Student's statistic at the 95% confidence level, and *error* is the allowable error fraction.

A value for n_1 is found iteratively by first using an estimate for n_1 and then comparing the estimated and actual values for n_2. The final value for n_1 should result in an estimated ratio value of n_1/n_2 which is less than or equal to the actual ratio value. An estimate for the boundary length per unit area LU is obtained as:

$$LU = \pi n_2 / 2L$$
(4.13)

where L is the total length of line transects, and n_2 follows Equation 4.12. Abeyta and Franklin (1998) applied line transect sampling to determine how closely the scrub vegetation stand boundaries derived from a segmentation algorithm corresponded to reference data in a California state park.

When both test and reference line coverages are available, it is of interest to assess line separation on the basis of complete line data. The measurement of mismatch between measured and true lines, however, is not straightforward. Unless individual points defining line segments are unambiguously identified on both measured and true maps, it is not simple to measure the epsilon distance. This is because there is no obvious basis on which to select a point on the true line as representing a point on the distorted line, especially when true lines are in reality curved. In many cases two measured versions of a line will have fewer vertices than the assumed true line because of the generalisation inherent in representation, leading to indeterminacy in matching vertices and hence non-uniqueness in evaluating the mismatch between lines, as shown in Figure 4.4.

A feasible method is to measure the gross misfit between the observed line and the assumed true line, as indicated by the shaded polygons caused by mismatch between the measured line and the true line in Figure 4.4. A probabilistic interpretation of epsilon error bands may be useful in this regard, and may possibly overcome the difficulty in estimating epsilon widths.

Thus, instead of directly calculating the misfits between the measured and true lines, Goodchild and Hunter (1997) suggested using the percentages of the measured line falling within bands of varying widths centred on the true line. Given an epsilon width ε, an error band can be generated around the reference or true line. From this, it will be possible to calculate the proportion of the length of the measured line lying within the band. The width ε corresponding to a certain proportion p is called the probabilistic epsilon band p-ε and can be derived through an iterative process (Goodchild and Hunter, 1997).

Such a method not only circumvents problems with vertex matching and outliers, but will also prove useful for the probabilistic quantification of positional uncertainties in objects more complex than points or lines. Areas may be represented as polygons that consist of a combination of topologically structured points and lines, or as contiguous groups of raster cells. Errors in the former case may be easily formulated through an extension of epsilon error bands. In the latter case error is expressed as a count of boundary cells, given a specific grid size, as shown in Figure 4.5, where boundary cells are shaded a darker grey.

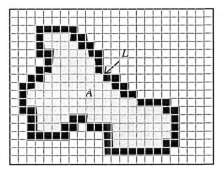

Figure 4.5 Gridding a polygon of area A and perimeter L.

The number of boundary cells for a gridded polygon, n_{cell}, has been shown to be estimated by:

$$n_{cell} = 2k_1\sqrt{\pi A}\big/k_2 l \tag{4.14}$$

where parameter k_1 is a shape factor, A is the polygon's areal extent, k_2 is a within-cell contortion parameter, and l is the average length of a random straight line laid across a grid cell (Crapper, 1980). The shape factor for a polygon may be quantified by comparison with other regular shapes, or formulated from basic parameters such as the polygonal area and perimeter (Frolov and Maling, 1969). As discussed by Goodchild and Moy (1976), the shape factor k_1 for a polygon with area A and perimeter L is defined to be:

$$k_1 = L\big/2\sqrt{\pi A} \tag{4.15}$$

An implicit assumption underlying the estimation of the number of boundary cells is that polygonal boundaries are well defined. This assumption may be reasonable for cultivated landscapes of intensive agriculture and pasture. However, many natural phenomena are not associated with sharp boundaries; for example, there are few such boundaries in forested areas, despite strong heterogeneity of species. Except for the convenience of land management practice, areal distributions are not readily modelled and manipulated as discrete objects. Instead, much spatial variability and heterogeneity have to be explored from a field-based perspective.

4.2.3 Describing Errors in Fields

As discussed in Chapter 2, many geographical distributions are more naturally and sensibly modelled as fields of continuous and categorical variables. Elevation as a kind of continuous variable forms an important basis for terrain modelling, especially when using digital techniques, while land cover is frequently quoted as a commonplace example of categorical mapping.

Errors due to measurement imprecision may be evaluated by repetitive sampling of the variable at a location, and possibly at a number of locations to reduce bias. The accuracy for a continuous variable can be assessed objectively based on the calculation of mean and standard error from a set of samples that compare measured values against reference data. Accuracy is then interpreted on the assumption of data normality, leading to the conclusion that the true value of the variable will not deviate more than twice the standard deviation from the mean value with a probability of about 95%.

Published DEMs are sometimes supplied with a report of the accuracy level as a reference for users. For example, the accuracy of the DEMs produced by the USGS is measured by RMSE. Suppose an RMSE of 50 cm is recorded for a spot height of 100 m; it is then inferred that the true height will be within the range 99 m to 101 m with a probability of about 95%. Hunter and Goodchild (1995) employed similar methods for the analysis of uncertainties involved in applications of DEMs to the estimation of the limits of a floodplain.

When a surface is expected to be essentially continuous, geostatistics can provide a useful theoretical and conceptual framework (Oliver and Webster, 1990). For instance, Kriging, as a spatial prediction and interpolation technique based on geostatistics, generates a variance surface as well as the estimation surface from a certain distribution of point observations or samples, with the former being a by-product for the latter. So, for every point in an estimated surface, a measure of uncertainty is readily available from the variance surface, as shown in Figure 4.6, where a hypothetical measured surface profile, $z(x)$, is accompanied by an interval of twice the standard error, $2std_z(x)$, which is grey toned, while dots stand for sampled locations.

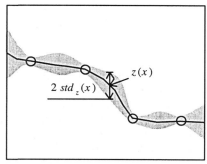

Figure 4.6 Profiles of a continuous variable $z(x)$ and standard error intervals $2std_z(x)$.

The interpretation of a variance surface may follow similar rules as seen in the case of positional errors, on the assumption of a normal distribution. Thus, the true surface value at location x, $Z(x)$, will fall within plus and minus one standard error $std_z(x)$ of the derived value $z(x)$ with a probability of about 68.3%, that is:

$$P(|Z(x) - z(x)| < std_z(x)) = 0.683 \tag{4.16}$$

For categorical variables, on the other hand, it is often sufficient to judge whether a labelling or classification is right or wrong. It is thus the practice to calculate the probability of a classification being right or wrong based on adequate sampling (Hord and Brooner, 1976; Hay, 1979).

As in remote sensing, the assessment of classification accuracy is based on a confusion or error matrix, which is derived after comparison of recorded classes against a reference source, often described as ground truth, based on a certain number and distribution of samples, by testing whether the classified data agree with the reference data. This proceeds by deciding on a suitable comparison unit, such as pixels in the case of digital images or polygons for photo-interpreted data. Reference data are usually provided by means of field survey, map data, or aerial photography. It is necessary that the two data sets register to one another and that they use mutually compatible classification systems with respect to the number and descriptions of classes and mapping units. Comparisons are then made based on overlay or by sampling. A superimposed overlay produces an overlaid map with a combination of classes from the test data and the reference data. A sampling process, on the other hand, generates a set of samples according to pre-selected sampling schemes.

It is straightforward to tabulate for each pixel or sample the predominant category shown on the reference map and the category as shown on the map or image to be evaluated. A summary of this tabulation forms the basis for the construction of the uncertainty or confusion matrix. For the purpose of illustration, an error matrix is provided in Table 4.1 with symbolic entries (e_{ij}, i,j=1,2,3), while Table 4.2 provides a numerical example. In Table 4.1 the dot subscript indicates summation over that index; for example, $e_{i.}$ is the sum over all columns of the entries in row i.

User's accuracy is the probability that a location labelled as category k (k=1,2,3) actually belongs to category k in the reference source ($e_{kk}/e_{k.}$), and is a measure of commission error; producer's accuracy is the probability that a location known to belong to category k is accurately labelled as category k ($e_{kk}/e_{.k}$), and is a measure of omission error. The overall classification accuracy (i.e., the observed proportion of agreement or per cent correctly classified) is calculated by dividing the sum of the diagonal entries e_{kk} in the error matrix by the total of all matrix elements. This overall classification accuracy is denoted by p_o, which is usually reported as the per cent correctly classified. For example in Table 4.2, p_o=69%.

The overall classification accuracy represents the probability that a randomly selected location is correctly classified. But it takes no account of chance agreement, because even a purely random assignment of class labels will result in a positive value. In other words, the overall classification accuracy tends to give an inflated index of classification accuracy, as Veregin (1995) described.

To remedy this shortcoming, the Kappa coefficient of agreement, Kp, has been recommended. As a quantitative measure of classification accuracy, Kp is calculated as:

$$Kp = (p_0 - p_e)/(1 - p_e) \qquad (4.17)$$

where p_0 is the observed proportion of agreement as explained previously, and p_e is the proportion of agreement that may be expected to occur by chance, calculated by summing the row and column marginals of the error matrix, that is:

$$p_e = \sum_{i=1}^{c} e_{i.}e_{.i}/e_{..} \qquad (4.18)$$

Table 4.1 An error matrix for classification accuracy assessment.

		Reference class			Row total	Marginal probability	User's accuracy
		U_1	U_2	U_3			
Map	U_1	e_{11}	e_{12}	e_{13}	$e_{1.}$	$e_{1.}/e_{..}$	$e_{11}/e_{1.}$
class	U_2	e_{21}	e_{22}	e_{23}	$e_{2.}$	$e_{2.}/e_{..}$	$e_{22}/e_{2.}$
	U_3	e_{31}	e_{32}	e_{33}	$e_{3.}$	$e_{3.}/e_{..}$	$e_{33}/e_{3.}$
Column total		$e_{.1}$	$e_{.2}$	$e_{.3}$	$e_{..}$		
Marginal probability		$e_{.1}/e_{..}$	$e_{.2}/e_{..}$	$e_{.3}/e_{..}$		1	
Producer's accuracy		$e_{11}/e_{.1}$	$e_{22}/e_{.2}$	$e_{33}/e_{.3}$			

Table 4.2 A numerical example of Table 4.1.

		Reference class			Row total	Marginal probability	User's accuracy
		Grass	Shrub	Wood			
Map	Grass	24	9	0	33	0.33	0.73
class	Shrub	10	25	5	40	0.40	0.63
	Wood	2	5	20	27	0.27	0.74
Column total		36	39	25	100		
Marginal probability		0.36	0.39	0.25		1	
Producer's accuracy		0.67	0.64	0.80			

For example, in Table 4.2, the chance agreement is 0.34, while the Kappa coefficient is 0.53, indicating that the accuracy of classification is 53% better than the accuracy that would have resulted from a random assignment. It has become a widely used measure of classification accuracy, because all elements in the classification error matrix contribute to its calculation, and because it compensates for chance agreement.

As has been seen so far in this section, every point of a field of categorical variables is associated with a discrete outcome, such as a nominal or an ordinal label in a classification system. Thus, an appropriate measure of uncertainty for a categorical variable at a point is the probability that the classification at that point is not correct. Such probabilistic measures may be obtained from repetitive sampling of individual locations, but it is more common to use an error matrix to derive a variety of measures such as the overall classification accuracy and the Kappa coefficient.

The measure of uncertainty mentioned above might suggest a spatially invariant level of uncertainty for each class. However, it is rarely the case that a coverage-wide classification can be reached with uniform accuracy or conceived as possessing an identical level of certainty. This is because many geographical categories exhibit significant spatial variation at a range of scales. But much spatial variability is suppressed during the forced categorisation necessary for incorporating discrete classes into spatial databases.

Suppose a class of grassland is found in two patches. It is not unusual to find that the growth of grass in these two patches is distinctively different. Even within each patch, conditions may not be exactly the same. Confronted with such spatial complexity, data analysts are likely to disagree with one another about the appropriate classification of land, and even a single human may find it difficult to achieve consistency in spatial categorisation. Moreover, such spatial heterogeneity is often recorded spectrally in remotely sensed images.

The spatial variability mentioned above suggests that natural occurrences or human interpretations of categorical variables should be considered as continuously varying to reflect the inherent variability in the underlying spatial categories and, hence, the conformity with which they are mapped. Thus, suppose c classes are possible. A suitable approach might be to view a set of categorical variables as forming a vector field of per-class probabilities $P(x)=(p_1(x),...,p_c(x))$, where elements $p_k(x)$ represent the probability of location x belonging to a candidate class k ($k=1,...,c$), and are required to sum to 1.

An illustration is provided in Figure 4.7, where an example of two classes U_1 and U_2 is shown conventionally in map form as (a) by mapping the most probable class at each point, while (b) and (c) depict, respectively, the spatially varying probabilities with which classes U_1 and U_2 occur. Thus, instead of a uniform level of uncertainty associated with each class, as suggested by the error matrix, the use of $p_k(x)$ facilitates representations of spatially varying categorical uncertainties. Mark and Csillag (1989) discussed the representation and derivation of probabilistic surfaces from epsilon error bands, an approach that may be relevant to the boundaries separating classes of natural resources, or ecological phenomena such as soil.

Cartographic conventions provide numerous ways of indicating overall levels of data accuracy. Indeed, there are numerous methods for describing map errors and making various statements about spatial accuracy on paper maps. It may be specified, for example, that the smallest map unit represented is about 10 m by 10 m on the ground, and that area boundaries are accurate to within +/- 15 m of their delineated position. It is sensible for topographic map users to interpret contoured elevation information with an accuracy level of half the contour interval, say 10 m. Most categorical maps contain hidden impurities or inclusions in uniformly coloured or shaded areas that amount to a substantial proportion of the areal extent assigned to the category. Similar descriptions may well be attached to the alphanumeric contents of digitised maps, or perhaps to the metadata, to assist potential users in judging each map's fitness for use.

Developments in GIS science and technologies have greatly enhanced the ways in which map data and geographical information in general are examined with respect to their inherent errors. This section has documented some of the key theories and techniques for describing geographical uncertainties in both objects and fields. It has been emphasised that uncertainties should be conceived as spatially varying. This is because variability in geographical forms and processes, and in human knowledge and understanding, and the physical limitations of spatial information technologies, along with various unknown factors, have all contributed to a view of geographical phenomena that is far less homogeneous than idealists might wish.

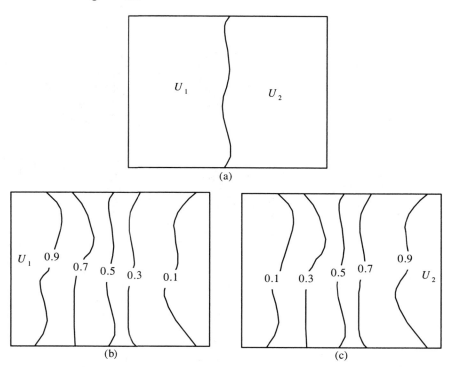

Figure 4.7 Representing categories U_1 and U_2: (a) mapping the most probable class; (b) and (c) mapping probability surfaces.

Even discrete objects are mapped with varying accuracies, depending on perceived usage, types of objects, and human and technical limitations in data acquisition. For instance, well-paved highways, which usually have distinctive shapes imposed as a result of engineering artefacts, lend themselves to easier positioning and less ambiguity than semi-natural footpaths or meandering streams, which may be difficult to identify on aerial images, or may be subject to substantial subjectivity or inconsistency in interpretation.

With the prevalence of uncertainty in heterogeneous geographical data, it is possible to extend the use of fields to the domain of discrete objects, whereby the positional errors associated with the measured positions of objects and boundaries are modelled as continuous vector fields. The advantages offered by fields in capturing spatial variability will be explored further elsewhere in this book.

4.3 UNCERTAINTY MODELLING

As has been discussed in the preceding section, uncertainties in both objects and fields are generally described using error statistics such as RMSE or probabilities of misclassification. A more important aspect of error description is the representation of spatial variation, which seeks to convey the spatial variability of uncertainties pertaining to mapped reality. Thus, geographical information users can be better informed than with non-spatial representations of errors.

Further analysis may be necessary, as spatial queries and decision-making often involve the evaluation of suitability given specific conditions, or of potential risks in taking or not taking certain actions. With imperfect data and knowledge, one may look to statistics and probability for possible answers. This amounts to modelling and assessing the significance of uncertainties as they propagate through GIS operations.

Error modelling and hypothesis testing based on normality assumptions have a long history in the scientific world. Variance propagation is a sound practice in experimental design and analysis, and the stochastic theory of variance propagation may also be used to model the propagation of errors in geo-processing. However, much spatial modelling involves heterogeneous data layers, and spatially varying uncertainties in source maps are likely to produce similar spatial variation in the uncertainties associated with output maps. The property of spatial heterogeneity raises questions about traditional statistical inference, which is based on the assumption that all cases in a sample are drawn randomly from the same population.

GIS communities have seen an impressive accumulation of useful work on modelling spatial errors. Examples involving continuously differentiable arithmetic operations have been demonstrated by Heuvelink (1998). The inputs for uncertainty modelling usually include surfaces of continuous variables and their corresponding surfaces of prediction errors, which may be obtained from Kriging or calculated from repetitive sampling. Errors associated with coefficients in spatial models are assumed to be the responsibility of experts in spatial modelling.

For categorical data as opposed to continuous data, Newcomer and Szajgin (1984) discussed the propagation of uncertainties during map overlay using probability theory. Their model accounts explicitly for the existence of correlations in errors by making use of conditional probability. Walsh *et al.* (1987) applied this model to quantify the uncertainty present in overlaid raster maps of land cover, slope angle, and soil type, with errors in the source layers measured by ground checking a set of sample cells.

The techniques mentioned above all attempt to quantify the propagation of errors in source data through arithmetic and logical GIS operations. However, many conventional approaches to error analysis assume that observations are drawn randomly and independently, an assumption that is invalid for a great deal of spatial data.

Consider the calculation of slope from elevation data. Slope gradients are defined as local estimates of the rate of rise and fall of terrain within certain neighbourhoods. As there are differences due to the manner in which elevation values at neighbouring locations are measured and derived, slope will be subject to various uncertainties due to errors in the initial elevation data. Further, errors in slope estimates will behave differently, depending on whether the neighbouring locations incorporated for slope calculation suffer from independent random errors, or whether these errors are correlated. In other words, error variance in slope estimates will depend not only on error variance in elevation measurements, but also on the spatial covariance of errors among neighbouring locations.

As noted earlier, spatial dependence suggests that things located closer together in space and time are more alike than things located further apart (Tobler, 1970). All spatially distributed phenomena display spatial dependence to some extent, and so do spatial errors. Geostatistical techniques have revealed that the phenomenon of spatial dependence is present in most if not all geographical phenomena. Interdependence among geographical variables is also commonly observed (Journel, 1996). It follows that uncertainties should be seen as spatially varied and correlated, similar to the way in which underlying geographical phenomena express themselves. Modelling spatial covariance and cross-covariance is thus crucial for realistic uncertainty analysis and error propagation.

Early efforts to model spatial distributions are largely developments of spatial auto-regressive methods (Goodchild, 1980). Geostatistics offers a selection of methods for modelling spatial errors, and the geostatistical technique of stochastic simulation is widely accepted in the geosciences. Stochastic simulation is used to generate equal-probable realisations of the spatially distributed variables present in a problem domain (Journel, 1996; Heuvelink, 1998).

Both unconditional and conditional modes for error simulation are useful. If sufficient evidence exists to support stationarity of spatial covariation of a variable over different regions, some regions are sampled but others not. One may wish to use spatial covariance models inferred from sampled regions for modelling unsampled regions, and unconditional simulation will be useful for such scenarios. On the other hand, when samples are available over a region, it is very sensible to generate realised fields conditional on samples of known values at specific locations.

Hypothetical examples, extended from the examples shown in Figure 4.6, are provided in Figure 4.8(a), where the true but usually unknown surface profile of a continuous variable $Z(x)$, drawn with solid line, is approximated with realised versions $z^1(x)$ and $z^2(x)$, which are drawn with darker and lighter dashed lines, respectively. The realisations shown in Figure 4.8(a) are conditional on the dotted sample locations.

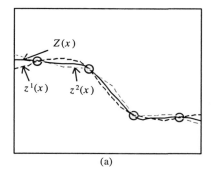
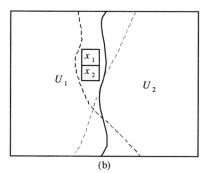

Figure 4.8 Simulating equal-probable realisations:
(a) a continuous variable $Z(x)$ seen in a profile; and (b) spatial categories U_1 and U_2.

Categorical maps may also be simulated using techniques adapted from those for continuous variables. Like the spatial dependence requirement for simulating fields of continuous variables, the observed spatial correlation in categorical occurrence needs to be honoured in simulated categorical maps. Illustrations built upon the examples shown in Figure 4.7 are provided in Figure 4.8(b), where two categories are shown with the true boundary as the solid line, and realised versions in darker and lighter grey dashed lines, respectively. Also indicated in Figure 4.8(b) are two neighbouring locations x_1 and x_2, whose values are closely related to each other in the realisations, as demanded by spatial dependence. Modelling joint occurrences of categories are at the core of simulation studies, which will be discussed in greater detail in Chapter 6. Goodchild *et al.* (1992) described a method for simulating categorical maps, which was based on raster data structures to avoid geometric interference with the probabilistic classification process.

Although spatial variability and dependence are easy to conceive for fields, it may seem a bit hasty to extend them to the case of discrete objects. Consider the process of map digitising. It would seem plausible to model the uncertainty of a line by modelling each point's accuracy, assuming that uncertainty in the line is entirely due to uncertainties in the points. However, lines and areas seem to be measured less accurately than individual points, in part as a consequence of generalisation, although lines and areas are represented as combinations of points (as polylines and polygons respectively) because of the discretisation necessary for digital mapping and analysis.

More seriously, points are captured subjectively during digitising, and the resulting positional errors of the points comprising the line are not independent, but tend to be correlated; this effect is commonly observed to be stronger when digitisers are operated in stream rather than point mode (Goodchild, 1991). The relationship between true and digitised lines, therefore, cannot be modelled as a series of independent errors or displacements of points.

An example is shown in Figure 4.9, where a highway drawn with parallels is seen intersecting a narrower road linking two buildings drawn as rectangles. In Figure 4.9, dots represent the points selected in digitising the road, and subsequently linked into the digitised line that is drawn with the darker dashes. Experimentation with map digitising has confirmed that serial or spatial correlations exist in the displacements of digitised points from the underlying line (Keefer *et al.*, 1991). Exceptions may occur with isolated artefacts such as the tics shown as small squares in Figure 4.9, which may be positioned independently.

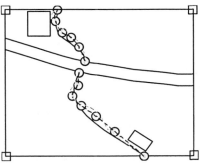

Figure 4.9 Lines digitised and simulated.

Once the existence of spatial correlation in the positional errors of digitised objects is recognised, the spatial covariance in positional errors along lines or in more complex objects may be quantified before being put to use in stochastic simulation. A hypothetical alternative realisation of the digitised road in Figure 4.9, using such covariance information, is shown as the lighter dashed line.

Thus far, stochastic simulation has been described at a conceptual level with illustrations. Further discussion may help to clarify the nature of geographical uncertainties and spatial errors in a broader sense. Error measures such as the standard deviation may be derived from multiple sampling, which, though very costly, serves to provide the raw data for calculation of means and standard deviations. This is only possible for well-defined variables like elevation, which, as a parameter of a visible and accessible landscape, can be measured several times.

For other variables, such as urban noise levels, atmospheric pressure, and population density, however, it is impossible to go back to collect more samples. Thus, multiple sampling cannot be used to measure the uncertainties involved in these variables. Besides, such variables are usually invisible and changeable, and their true values are known only by more detailed sampling and more exact measurement (Carter, 1988). These extra samples are used to detect gross aberrations, but provide no basis to confirm detail. Further, any interpolation from samples of such variables is subjective, because it is only possible to judge whether an interpolated surface is plausible or absurd based on conceptual and theoretical expectations, and impossible to confirm whether the data are right or wrong.

With this understanding, it is important to realise that much of geostatistics is built upon modelling, and that stationarity is an assumption made for a model rather than an empirical observation about the geographical phenomena being modelled. Thus, stochastic simulation is really an invaluable asset to enhancing geographical information with additional information about uncertainty. If spatial covariance and cross-covariance are quantifiable, it becomes possible to model any spatially distributed phenomena, be they continuous or categorical in nature. The chapters that follow will explore these techniques further with real-world examples, in order to provide realistic perspectives on spatial inaccuracy and better understanding of geographical phenomena that are uncertain to varying degrees and in various ways.

4.4 MORE UNCERTAINTY

Concepts and theories of probability have strengthened the quantitative analysis of geographical data, and their relevance to scientific research as well as to daily life has been widely accepted. The classical concept of probability, developed originally in connection with games of chance, is based on the proportion of outcomes of an experiment or trial that yield a particular result. The concepts of sample space, sets, and events provide a rigorous basis for probability. In the early eighteenth century, the French naturalist George de Buffon calculated the probability that a very fine needle of length a, which is thrown random on a board ruled with equidistant parallel lines spaced b apart, will intersect one of the lines; the result is given by $2a/(\pi b)$. Thus, the probability of an event is interpreted as the proportion of times that similar events will occur in the long run, although there is no guarantee of the outcome on any particular occasion. This so-called *frequentist* interpretation underlies the classical concept of probability (Kolmogorov, 1956). An immediate consequence is that the exact value of probability is never known but can nevertheless be estimated with reasonable accuracy, in particular if one has the luxury of large samples.

Classical probabilities always refer to what happens in similar situations, such as the flip of coin. When uncertainties exist with regard to single events, associated probabilities must be obtained by the aggregation of large numbers of identical events. In situations in which little is known directly, one has to resort to considering a combination of collateral information, educated guesswork, and other subjectivity (Good, 1950; Kyburg and Smokler, 1964; Jeffreys, 1983). A common example is the expression of odds in gambling, or in the informal prediction of rain events. Odds are often related to the ratio of the amount a person feels that he or she should gain in return for risking a certain amount in a given situation. An odds of $a{:}b$ may be converted into a probability $p=a/(a+b)$. Since odds are likely to be differently assessed by different people, the practical value of subjective or personal probability is clearly open to question. Although useful assessments can be given of subjective probability by analysis of a person's experience in the evaluation of chance, past performance, and familiarity with the circumstances, the human mind has proven itself relatively inept at estimating odds throughout history (Freund, 1973). This reflects the fundamentally intuitive rather than logic nature of the subjectivity involved in probabilistic reasoning, which is echoed in the superstitions associated with gambling.

In soil survey, measurements of chemical and physical properties often comprise the quantitative description of a soil profile, as discussed by Mark and Csillag (1989). Usually, a set of rules would assign each location to one of a prescribed number of mutually exclusive and collectively exhaustive soil classes. Since there are errors in soil sampling data, the fact that the class assigned at a location is a deterministic function of observed quantities might indicate that probabilistic approaches are all that one needs to cope with uncertainty in soil classification. However, the human mind is an indispensable contributor to geographical data and analysis. In spatial classification, humans tend to divide real-world phenomena into categories in ways that are relatively consistent among individual minds (Johnston, 1968). In general, spatial problem formulation and solution have to consider human cognition, for which subjective interpretation and the use of probability may not provide an adequate conceptual framework.

In land evaluation, a combination of qualitative and quantitative values is used to define specific land-quality and suitability classes through logic operators. Ordinal classes such as grades of suitability are commonly defined by specifying the numerical ranges of a certain number of key properties that an individual land unit must meet to qualify for certain classes. While it is easy to tell the difference between extremely suitable and unsuitable land units, differentiation between less distinctively defined categories is clearly more difficult.

Numerous examples exist, such as the gradation from rural to urban areas, which is seldom clear-cut. While there may be general agreement about the core concepts of particular spatial objects and categories, there is more likely to be uncertainty regarding the boundaries of their spatial and semantic extents. Thus, in practice, it is far from easy to draw discrete breaks between the spatial entities of metropolitan areas, towns, and villages. Similar problems come to light when data analysts try to define exactly how wet a land patch needs to be to be labelled as wetland, and where to place wetland boundaries. Many spatial categories are not as exactly defined as one might wish.

Traditional Boolean set theory, in which set assignments are crisp, fails to represent such vague entities or categories; that is, set characteristic functions assume values of 0 or 1. Boolean set logic implies that all universes of entities can be partitioned into exhaustive and mutually exclusive sets. As human and physical phenomena involve much vagueness and subjectivity, exemplified by spatial choice sets and subjective preference relations, it is doubtful that crisp sets are capable of providing an adequate account (Pipkin, 1978; Leung, 1987).

In close association with vagueness is ambiguity. Ambiguity in cognition, with respect to conceptual and linguistic categories, arises from inconsistency in evaluating the set structures embedded in algebraic and probabilistic models of choice. Typically, these ambiguities have been attributed to human cognitive limitations, and handled essentially as statistical error. A better procedure is to account for these problems as shortcomings in the concept of set itself. However, Black (1937) distanced ambiguity from the umbrella of vagueness, and discussed the tricky relations between the notions of generality and vagueness, which were researched by Zhang (1998) in linguistic studies.

Put simply, vagueness includes fuzziness in the classes themselves as well as cognitive inexactness in rating. Subjective estimates of physical distance or travel time are frequently rounded to simple cognitive units of miles or minutes. The inexactness of such predicates as 'reasonably close' for locations x_1 and x_2 may be formulated by an exponential function for closeness with Euclidean distance acting as the exponent prefixed with a negative sign, that is, $\exp(-\|x_1-x_2\|)$. Thus, it is natural to approach vagueness directly by employing graduated set membership explicitly. Fuzzy sets, initially proposed by Zadeh (1965), seem to be capable of unifying apparently disparate concepts such as value, preference, constraint, goal, and decision into grades in the unit interval [0,1].

Let U be a universe of discourse, whose generic elements are denoted by $x:U=\{x\}$. The membership in a classical set U_1 of U is often viewed as a characteristic function from U to $\{0,1\}$ such that $U_1(x)=1$ if, and only if, $x \in U_1$. A fuzzy set U_1 in U, on the other hand, is characterised by a fuzzy membership function $\mu_1(x)$, which associates with each x a real number ranging from 0 to 1. The fuzzy set U_1 can be designated as:

$$U_1 = \left\{ \left(x, \mu_1(x)\right) \middle| \ x \in U \right\} \tag{4.19}$$

where the value of $\mu_1(x)$ at x represents the degree of membership of x in U_1, and is commonly termed a fuzzy membership value. The closer the fuzzy membership value is to 1, the higher the degree of membership of x in U_1. The fuzzy membership function $\mu_1(x)$ may be probabilistically estimated, but may also be based on subjective assessment.

Despite the empirical and methodological appeal of fuzzy sets in spatial cognition and decision-making, substantial problems remain to be overcome in implementation. Though the general characteristics of fuzzy measures may be established empirically, it is clearly impossible to identify such functions precisely. In many practical situations, *a priori* knowledge about the characteristic function of a fuzzy set is not sufficient to construct an optimal estimate. Thus, in many instances, one is forced to resort to a heuristic rule for estimating fuzzy membership functions, whose appropriateness can only be verified by experimentation. Indeed, a very noticeable feature of applications using fuzzy sets is the ad hoc and even apparently casual way in which the fuzzy membership functions are specified. On the contrary, probability theory seems to be more firmly grounded with both frequentist and subjective types of interpretations.

The differences between fuzzy sets and probability measures should not be confused by the superficial and numerical similarity between them, as both are valued in the unit interval. Though distinctions between fuzzy logic and probability theory are increasingly emphasised in the literature, their practical implications are less clear. It may be possible to formulate psychologically based but purely probabilistic accounts of individual preference in spatial choice. Still, fuzzy logic may provide a conceptually more natural account of subjective inexactness (Pipkin, 1978), though intuitive and personal elements are acknowledged in subjective interpretation of probability (Fishburn, 1986). Despite the fact that fuzzy memberships are themselves phrased in the usual mathematics of exact numbers and Boolean logic, fuzzy memberships described thereby are, at best, approximations to inherently inexact concepts and linguistic variables, a topic revisited by Nguyen (1997) and Zadeh (1997).

Despite these more tricky aspects of uncertainty reasoning with probabilistic and fuzzy paradigms, both are clearly useful and should be seen as complementary to each other in representing and handling uncertainty. The next three chapters aim to explore both methodologies in geographical information processing, where the spatial distributions of quantities, qualities, and objects are studied from the perspectives of spatial variability and spatial dependence.

CHAPTER 5

Uncertainty in Continuous Variables

5.1 INTRODUCTION

Fields have been considered in Chapter 2 as being important models for spatial data, as they are widely recognised as being flexible and powerful for capturing spatial variability and heterogeneity in geographical reality, and are thus preferred for purposes of uncertainty handling (Goodchild, 1989; Kemp, 1997; Vckovski, 1998). In the field domain, variables may be discrete- or continuous-valued. While both categorical and continuous variables are relevant to fields, the latter seem be closer to the classical notion of continuous variation, and often serve to bridge the gap between idealised discreteness and realistic continua. In light of this, it is sensible to start the exploration of geographical uncertainties from the domain of continuous fields. This chapter aims to provide an account of field-based methodologies for handling spatial uncertainties, backed up by experimentation with real data.

A commonly known example of continuous variables is terrain elevation, whose spatial distribution above some arbitrary datum in a landscape is represented as structured arrays of numbers known as digital elevation models (DEMs) or digital terrain models (DTMs) in a GIS environment (Mark, 1978; Jenson and Domingue, 1988; Clarke, 1995). Discussion of fields and their uncertainties can thus be usefully referred to elevation data. Accuracy of elevation data has been an important theme for research among photogrammetrists and mapping scientists (Frederiksen et al., 1985; McCullagh, 1988). However, despite this early work a full treatment of accuracies remains on the research agenda, due to the complex interactions between different types of errors, spatial irregularities, and other matters of complexity, as will be examined in this chapter. Thus, discussion of DEM-related uncertainty is an endeavour of great significance.

Elevation may be sampled at discrete points or as the average elevation over a specified segment of the landscape. Slope or slope gradient has always been an important and widely used topographic variable in geomorphological and ecological studies, and its derivation from DEMs is increasingly computerised. Many land capability classification systems utilise slope as the primary basis for ascribing capability classes, together with other factors such as soil depth, as reviewed by Moore et al. (1993).

The specific context of uncertainty has to be identified from the start. If DEMs are created from photogrammetrically sampled elevation data, there will be uncertainties due to the complex interactions of image sources, the photogrammetric equipment, the data reduction techniques adopted, and the operators who perform height sampling, as has been discussed in Chapter 3. The U.S. Geological Survey (USGS, 1997) provides a categorisation of DEM data errors in line with statistical conventions. Blunders are recognised as extreme data values; systematic errors are often patterned, with constant magnitude or sign across a large section of the DEM; and random errors are, by definition, those that remain after removing the previous two.

Once the context of uncertainty is established, the description of uncertainties follows. For digital elevation data, statistics such as root mean squared error (RMSE) are useful measures, and are often computed from ground control points (GCPs) located on a specific DEM layer. For example, the USGS DEM data of level-3 accuracy, produced from linear interpolation of digital line graph (DLG) data using digitised contours and drainage networks, are constrained to a maximum error of 1/3 contour interval. For USGS DEM data of level-1 accuracy, it is required that an RMSE of 7 m be maintained for 90% of the grid cells (USGS, 1997).

For a field variable, error measures alone are of limited use for spatial decision-making. Error modelling is more important because it provides a workable mechanism for propagating the errors present in the source data as they are transferred through GIS operations and accumulated in outputs. But modelling spatial uncertainty is by no means trivial, due to the existence of spatial dependence in and among data layers. If spatial dependence is assumed absent, then uncertainty modelling would simply be a matter of averaging for a global statistic, such as RMSE, or summation of the variances involved in an algebraic map overlay across data layers. But many geographical variables are cross-correlated as well as spatially auto-correlated, making the task of modelling spatial uncertainty extremely demanding.

Consider the accuracy of slope data calculated from a USGS square gridded DEM data set with a point spacing of 30 m and with a standard deviation (σ) of 7 m. Suppose slope is calculated as the difference in elevation at adjacent locations divided by the grid spacing. Applying variance propagation leads to a meaningful assessment of slope accuracy, assuming that all relevant parameters are specified, including the standard deviation σ and the spatial correlation of elevation errors at adjacent locations (ρ). Goodchild (1995) shows that the standard deviation in slope is given by the square root of $2\sigma^2(1-\rho)$, divided by the point spacing (in effect, the linear size of square pixels). If there is zero correlation ($\rho=0$), the standard deviation in slope would be no less than 33%, far too high for slope estimated in this way to be of practical value, while in the case of perfect correlation ($\rho=1$) slope estimates would be error-free. In reality, such slope data are sufficiently accurate to be useful, suggesting that ρ is substantially greater than zero.

It is possible to perform simple error analysis when the process for deriving an output variable from input data can be modelled as a simple algebraic relationship, and when the spatial variability of errors is properly accommodated. Many GIS applications, however, are sufficiently complex to make this analytic approach impractical. Instead, stochastic simulation, often known as Monte Carlo simulation, offers more flexibility and tractability for modelling uncertain systems (Ripley, 1992). The key to this is to devise stochastic methods for drawing equally probable samples of an error-contaminated field from distributions of all possible samples, and to implement them digitally. A good example for this is the Gaussian or normal distribution density function, which can be simulated with a computer by generating pseudo-random numbers and processing them using specialised algorithms.

Based on a suitable model, both data producers and data users are able to assess the fitness of a particular data product for a certain purpose by checking relevant statistics based on a chosen stochastic model. Indeed, researchers have made significant progress in modelling uncertainties in spatial data generally and in DEMs and their derivative data products specifically using this strategy (Goodchild, 1995; Fisher, 1998; Heuvelink, 1998; Liu and Jezek, 1999; Holmes *et al.*, 2000).

In practice, the data available for a particular application are likely to be heterogeneous, derived from many sources, and with complex properties. Individual spot heights can be surveyed directly in the field, and used to build a triangulated irregular network (TIN) (Peucker and Douglas, 1975). Digital elevation data may be derived by digitising contour lines from existing topographic maps, using the process of raster scanning and vectorisation, or manually by using a flat-bed digitiser and software packages. Photogrammetric techniques are available to produce DEMs from stereo-photo pairs, followed by either on-line or off-line processing. Stereo images, either digitised or acquired already in digital format, can be used for the semi-automatic production of digital elevation data based on digital photogrammetric workstations.

Elevation data that are ground surveyed or visually sampled from reconstituted image pairs are widely believed to be of higher reliability and accuracy than those automatically calculated from digital photogrammetric models. However, the point-by-point sampling used in the former case is an expensive and tedious process. Besides, the accuracy of photogrammetric data measured manually is not always ideal. The accuracy of digital photogrammetric mapping has been shown to be subject to the unreliability of the image matching process, and its resulting value is in turn the result of the combined interaction of image quality, algorithmic robustness, and scene complexity. Claims of high accuracy for high-resolution data generated from digital photogrammetry should not be rejected out of hand, as such data can be valuable primary sources of terrain information. Among other issues, there is the problem of complex interactions between terrain and non-terrain image segments in digital photogrammetry, since the latter provide the basis for accurate matching of conjugate image points and thus for the reliable reduction of elevation data. If non-terrain segments can be identified independently from synoptically imaged landscapes, the value of digital photogrammetry in terms of efficiency and reliability can be much higher.

Digital photogrammetry must deal simultaneously with issues of reality, modelling, measurement, and data, all of which have been discussed in earlier chapters. High-resolution images record measurements from spectrally visible terrain and its contextual landscape—a reality characterised by impurity and non-homogeneity. On the other hand, location-specific data surveyed on the ground or generated by photogrammetric measurements at human discretion are affected by implicit models of the terrain, and are often sparse samples from an abstracted population, and thus more remote from reality. The implication is that any meaningful exploration of geographical uncertainty must pay special attention to the complex interactions of heterogeneous data, which often have incompatible interpretations.

Given such concerns, geographical modellers would be well advised to seek flexibility from a broader territory of spatial statistics that goes well beyond traditional geostatistics and, in particular, from the Markovian legacy of local specification and Monte Carlo simulation. Markov chains are potentially of great significance for spatial modelling, allowing spatial complexity and data irregularities to be addressed by well-designed simulators (Bartlett, 1967; Cressie, 1993; Brémaud, 1999). Uncertainty modellers should benefit from many of the recent developments in spatial statistical methodologies.

This chapter will discuss various forms of errors in continuous variables, and will address the probabilistic estimation of uncertain distributions in terms of confidence intervals, for which indicator Kriging plays an essential role. The importance of accommodating spatial dependence within and between the data layers involved will be addressed in the course of modelling errors and their propagation. Descriptions are given of geostatistics and Markovian spatial statistics, which are seen as complementary methodologies with the potential to enhance the exploration of spatially correlated variations.

Using real elevation data derived from ground-based surveying, together with analytical and digital photogrammetry over urbanised areas, empirical studies were carried out to test the performances of Kriging, the analytical propagation of elevation errors, stochastic simulation for spatial mapping, and uncertainty modelling of slope data derived from elevation data. This is attractive as a data-driven and practical strategy that also offers verifiability from co-registered data of greater accuracy. A practical discussion of data analysis relevant to digital elevation data and slope modelling will be followed by a discussion of spatial statistical methods for handling uncertain geographical information.

5.2 FIELDS OF ERRORS AND PROBABILITIES

It was established previously in this book that errors and uncertainties accumulate in geographical data in various forms and disguises. Geographical data are intended to represent underlying geographical phenomena within specified tolerances that reflect specific requirements. Fields of continuous quantities are often mapped for a range of purposes, using sampled data followed by spatial interpolation. DEMs are routinely produced by using a range of techniques, to depict terrain that is physically changing and increasingly cultivated by humans.

Accuracy is a major concern for fields, in addition to issues of acquisition, storage, representation, and visualisation. For DEMs, research efforts include the Fourier analysis of topographic surfaces (Balce, 1987), the mathematical modelling of DEM accuracy (Li, 1994), and auto-regressive modelling of uncertainties in DEMs and their derivative variables such as slope and aspect (Hunter and Goodchild, 1997).

Recall the assumption made for ordinary Kriging and, in particular, the formulation of a stationary variable $Z(x)$ in Equation 2.11. A variable $Z(x)$ is viewed as the sum of a deterministic term $m_Z(x)$ and errors of spatially correlated and independent components. Assume that one source of reference data is available, which can be taken as constituting the true variable $Z(x)$. Denote the approximation of variable $Z(x)$ derived from measurement or calculation as $z(x)$. Since error is defined in Chapters 1 and 4 as the difference between reference and measurement, it might be possible to view the underlying field $Z(x)$ as the sum of its approximation $z(x)$ and an error term $\varepsilon_Z(x)$, as shown below:

$$Z(x) = z(x) + \varepsilon_z(x) \tag{5.1}$$

for which Figure 5.1 gives straightforward illustrations. In Figure 5.1(a), the grey blobs indicate sample points, with the solid and dashed profiles standing for the measured and the true yet unknown surfaces, respectively. The error surface is displayed along with its zero level in Figure 5.1 (b).

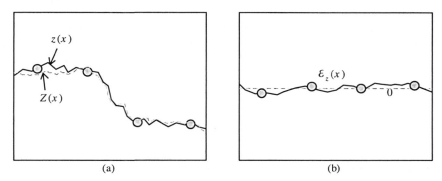

(a)   (b)

Figure 5.1 Mapping an uncertain field $Z(x)$: (a) measured version $z(x)$; and (b) error field $\varepsilon_z(x)$.

Typically, the accuracies of field quantities are expressed with error measures including RMSEs, which are evaluated over the spatial extent of the field under study. However, non-spatial error measures such as RMSEs say little about the spatial correlations present in the errors $\varepsilon_z(x)$, which will often be significantly non-zero. This is reminiscent of the process of regression-based spatial interpolation, where residual errors are assumed to draw from a population of zero mean and common variance, with zero covariance implying independence among data locations.

Geostatistical approaches to modelling fields support greater spatial variability. Kriging variance, calculated from Equation 2.31, is generated as a by-product of a Kriging procedure for estimating stationary random variables. It may be transferred to standard error for visualisation of spatially varying accuracy in the estimated field (Clarke, 1995). Kriging variance and, more generally, the estimation variance of linear geostatistics, $\sigma_{Z(x)}^2 = E[Z(x)-z(x)]^2$, are independent of the data values used to drive a specific Kriging system, once a suitable semivariogram model is identified from the data set. For this reason, Kriging variance is often considered merely as an index of data configuration used to build a norm for minimisation, rather than as a measure of local accuracy for the estimated field $z^*(x)$ (Journel, 1986).

It sounds intuitively more reliable to base error analysis on an evaluation of the specific distribution of data locations against an assumed reference. The geostatistical treatment of such location-specific errors should provide a more objective understanding of their spatial variability by taking account also of their spatial correlation.

The possible existence of trends in stationary random fields presents ambiguity as to whether to use universal Kriging or ordinary Kriging. In many cases, the formulation of a random variable as the sum of a measured component and an error term, as in Equation 5.1, resolves this issue by equating trends in the underlying fields to the measured versions, assuming no major systematic biases in measurement. The mapping of error fields becomes straightforward, and facilitates the geostatistical analysis of error-corrupted fields (compare Figure 5.1, where the error field $\varepsilon_z(x)$ may be subjected to geostatistical analysis).

This approach is valuable for digital elevation modelling. Despite their familiarity to GIS users, elevation data have only rarely been analysed within a geostatistical framework (Carter, 1992), despite the recent application of geostatistics to terrain modelling (Lane, 2000). However, if it is possible to derive errors at certain locations, it is generally reasonable to assert stationarity in fields of errors rather than in the variables under study, and to analyse them geostatistically.

While errors can be dealt with specifically as above, mapping spatial quantities with improved accuracy yet at decreased cost may be more sensible. As mentioned in the introductory section, GIS communities are no longer lacking in spatial data. Rather, abundant but heterogeneous data with unequal and possibly incompatible accuracies are often accessible for spatial problem-solving. Thus, instead of testing one source of data against another, it may be possible to improve accuracy by conflating data of varying accuracies into a single data set with an optimal combination of location-specific accuracy and overall variability. This may be addressed using the technique of co-Kriging (Myers, 1982; Kyriakidis *et al.*, 1999).

Suppose $z1(x_s)$ $(s=1,\ldots,n_1)$ and $z2(x_s)$ $(s=n_1+1,\ldots,n_1+n_2)$ are the so-called primary and secondary data sampled from the domains of fields $Z1(x)$ and $Z2(x)$, both pertaining to the unknown field $Z(x)$. When the primary and secondary sources are co-located, that is, when it is possible to ascertain that $n_1=n_2=n$ and $x_s=x_{n+s}$ for all s, quantification of spatial covariances cov_{Z1} and cov_{Z2} for the primary and secondary variables and their cross-covariance cov_{Z1Z2} will be straightforward using Equations 2.14 and 2.19 respectively. The cross-semivariogram γ_{Z1Z2} is related to the cross-correlogram ρ_{Z1Z2} via:

$$\gamma_{Z1Z2}(h) = \sqrt{\mathrm{cov}_{Z1}(0)\,\mathrm{cov}_{Z2}(0)}\,(\rho_{Z1Z2}(0) - \rho_{Z1Z2}(h)) \qquad (5.2)$$

When primary and secondary data do not coincide or overlap, as is often the case in practice, $\rho_{Z1Z2}(h)$ may be calculated from sample data for lags h other than zero, but $\rho_{Z1Z2}(0)$ needs to be inferred from the cross-correlogram $\rho_{Z1Z2}(h)$ by extrapolating to zero lag. Alternatively, pseudo cross-semivariograms can be defined to permit the matrix formulation of co-Kriging systems in lieu of conventional cross-semivariograms (Myers, 1991).

The technique of co-Kriging for interpolating the Z field is basically to search for the best linear estimator, denoted by $z(x)^*$, which is a weighted sum of local data:

$$z(x)^* = \sum_{s=1}^{n_1} \lambda_s z1(x_s) + \sum_{s=n_1+1}^{n_1+n_2} \lambda_s z2(x_s) \qquad (5.3)$$

where λ_s stands for the weight attached to the sth data point. The estimation is subject to a selection of conditions:

$$\sum_{s=1}^{n_1} \lambda_s = 1 \text{ and } \sum_{s=n_1+1}^{n_1+n_2} \lambda_s = 0 \qquad (5.4a)$$

or:

$$\sum_{s=1}^{n_1+n_2} \lambda_s = 1 \qquad\qquad (5.4\text{b})$$

except for simple co-Kriging where no constraints are imposed on the weights.

Solution of a co-Kriging system is not fundamentally different from that of ordinary Kriging (Myers, 1982; Cressie, 1993), and thus is not further elaborated here. A special case occurs when co-located grid data are available, providing values at each of the grid nodes to be estimated. Co-Kriging may be modified to exploit the specifics of data configuration (Deutsch and Journel, 1998), and examples are numerous. A ground-surveyed data set of crop yield may be available and taken as the primary data, while remotely sensed images recording vegetation and biomass radiance may be incorporated as secondary data. Kyriakidis *et al.* (1999) applied co-Kriging to conflate one set of elevation data of higher accuracy but sparser sampling, with another of lower accuracy but denser sampling. This geostatistical approach was remarkably effective for correlation analysis and spatial modelling, where data of substantially different accuracies exist, as is often the case. A hypothetical example is given in Figure 5.2, where primary and secondary data are shown as dots and grey thick lines, respectively, with the Kriged surface profile shown by the thin darker line.

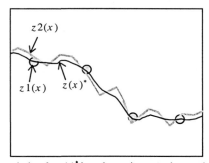

Figure 5.2 Co-Kriging solution for $z(x)^*$ based on primary and secondary data $z1(x)$ and $z2(x)$.

Mapping error fields is made possible by the existence of certain sources of higher accuracy that can be used to benchmark the specific data sets under scrutiny. More intriguing situations arise due to the complex interaction of the measured and noise-like components of random variables. In most cases, errors in a strict sense simply do not exist in reality, as many spatially distributed things exist in their entirety, and it is consequently impossible to conceive of a larger population from which the observed data are drawn. Kriging variance alone cannot provide objective estimates of confidence intervals for the underlying fields, unless multivariate Gaussian distributions are assumed, which are often unrealistic.

As the evaluation of confidence intervals is more important than the description of error fields, since the former provide critical information about uncertainty in the underlying variables, methods for mapping the variability in a particular variable without stringent distribution assumptions are welcome. To estimate confidence intervals, consider discretising the underlying variable of known range $[z_{min}, z_{max}]$ with c increasing cut-off values z_k ($k=1,\dots,c$). For each cut-off value z_k, it is possible to define an indicator transform $i(z_k;x)$ for any location x

such that:
$$i(z_k;x) = \{1 \text{ if } z(x) \le z_k; 0 \text{ if } z(x) > z_k; k = 1,...,c\} \tag{5.5}$$

The indicator transforms thus obtained can be conceptualised as random variables. The stationary mean of the indicator variable is the cumulative distribution function (cdf) of the random variable $Z(x)$ itself; this is easily proven for a cut-off value z_k ($k=1,...,c$):

$$E[I(z_k;x)] = 1 \times \text{prob}\{Z(x) \le z_k\} + 0 \times \text{prob}\{Z(x) > z_k\}$$
$$= \text{prob}\{Z(x) \le z_k\} = F(z_k;x) \tag{5.6}$$

The indicator transform of sample point x_s is denoted by $i(z_k;x_s)$, where z_k stands for the cut-off values ($k=1,...,c$). The experimental semivariogram can be calculated from the sample data:

$$\gamma(z_k;\hat{h}) = \frac{1}{2M(h)} \sum_{x_{s1}-x_{s2}=h} (i(z_k;x_{s1}) - i(z_k;x_{s2}))^2 \tag{5.7}$$

where $M(h)$ is the number of pairs of observation points $s1,s2$ separated by vector lag h, and where index $s1$ denotes a point in the primary set and $s2$ denotes a point in the secondary set. By changing h, a set of values is obtained, from which the experimental semivariogram is constructed.

Suppose a set of sample measurements exist, $\{z(x_s), s=1,...,n\}$. Given a cut-off value z_k, these samples can be easily transformed into indicator data $\{i(z_k;x_s), s=1,...,n\}$. Further suppose that an experimental semivariogram has been calculated by using Equation 2.13, and has been subsequently fitted using a suitable model. Indicator Kriging provides a least-squares estimate of the conditional posterior cumulative distribution function at the cut-off value z_k:

$$i(z_k;x)^* = E[I(z_k;x)|\text{data}]^* = \text{prob}\{Z(x) \le z_k|\text{data}\}^*$$
$$= \sum_{s=1}^{n} \lambda(z_k;x_s)i(z_k;x_s) + \left(1 - \sum_{s=1}^{n} \lambda(z_k;x_s)\right)F(z_k) \tag{5.8}$$

where the weights corresponding to a particular cut-off value z_k (i.e., $\lambda(z_k;x_s)$) are obtained by solving a simple Kriging system, and $F(z_k)$ is the global cumulative distribution function corresponding to the cut-off value z_k.

An important property of indicator Kriging is that, when the cut-off values are varied, the sequence of indicator data can be seen as an approximation to a distribution function, leading to a non-parametric solution to probabilistic mapping. An illustration is provided in Figure 5.3, where the posterior cumulative distribution function at a location x is estimated for a series of cut-off values z_k, $k = 1,...,c$, based on samples shown as grey-scaled dots.

For instance, the posterior cdf of the type estimated by Equation 5.8 will provide a clue to the characterisation of uncertainty about the unknown value $z(x)$ of the model random variable $Z(x)$. From such a posterior cdf, confidence intervals are derived directly and are independent of the particular estimate $z(x)^*$:

$$\text{prob}\{Z(x) \in [q_p(x), q_{1-p}(x)] | \text{data}\} = 1 - 2p \tag{5.9}$$

with $q_p(x)$ such that $F(q_p(x) | \text{data}) = p \in [0.0,0.5]$, being the pth quantile associated with the posterior cdf; similarly $q_{1-p}(x)$ is the $(1-p)$th quantile.

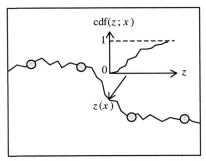

Figure 5.3 Probabilistic mapping $z(x)$ by indicator Kriging.

Various possible estimates for the unknown value $z(x)$ can be derived. One such estimate is the conditional expectation $z_E(x)^*$:

$$z_E(x)^* = E[Z(x)|\text{data}] = \int_{-\infty}^{\infty} z \, dF(z;x|\text{data})$$

$$\approx \sum_{k=1}^{c+1} z_k' \left((F(z_k;x|\text{data}) - (F(z_{k-1};x|\text{data}) \right) \tag{5.10}$$

where z_k ($k=1,...,c$) are the c cut-off values, $z_0 = z_{\min}$, $z_{c+1} = z_{\max}$ are the minimum and maximum of the z data, and z_k' is the conditional mean value within each class $(z_{k-1}, z_k]$. The conditional expectation thus obtained minimises the conditional estimation variance $E[(Z(x) - z(x)^*)^2 \mid \text{data}]$.

Environmental decision-making often involves setting specific thresholds for certain hazardous contaminant or pollutant concentrations, and figuring out the likelihood that contamination or pollution exceeds the prescribed thresholds. Indicator Kriging is readily applied to such situations, estimating the conditional or *a posteriori* probability distribution that a variable such as contamination concentration does not exceed a certain threshold or cut-off value. No particular assumption is made regarding the distribution of the underlying variable Z, and indicator methods have been usefully explored in environmental studies (Curran, 1988; Cressie, 1993; Rossi *et al.*, 1994).

In addition to sample data that are assumed to be perfectly known, indicator Kriging can also incorporate data that take the form of inequalities such as $z(x_a) \in (a_a, b_a)$, or $z(x_a) \in (-\infty, b_a)$ or $z(x_a) \in (a_a, +\infty)$. Suppose an inequality constraint exists for a local value such that $z(x_a) \in (a_a, b_a)$, then it is possible to define indicator data $j(z;x_a)$:

$$j(z; x_\alpha) = \{0 \text{ if } z \le a_\alpha; \text{undefined if } z \in (a_\alpha, b_\alpha); 1 \text{ if } z > b_\alpha\} \tag{5.11}$$

Ancillary data are sometimes available to provide prior probabilities about the value $z(x)$ in a global form $\text{prob}\{Z(x) \le z \mid \text{ancillary data}\}$ or in a local form $\text{prob}\{Z(x_a) \le z \mid \text{ancillary data}\}$. If so, coding of such probabilistic data is straightforward, as the indicator Kriging formalism lends itself easily to the probabilistic modelling set out in Equation 5.8.

One possible difficulty of indicator Kriging is the so-called order relation problem, which occurs when an estimated distribution function is decreasing, or has values outside of the interval [0,1]. Order relation problems happen because each indicator Kriging system corresponding to a particular cut-off value is solved independently from those at all other cut-off values, implying no guarantee that the solutions taken together will be valid as a probability function.

One method for resolving the order relation problem is to combine the indicator Kriging systems for all cut-off values, but this would incur an enormous expense. Probability Kriging is a method for reducing the risk of order relation problems by utilising more of the information present in the bivariate distribution of $z(x)$ and $z(x+h)$, and combining both z and indicator data in the solution, as formulated in indicator co-Kriging by Sullivan (1984).

There is, however, a persistent issue of determining the optimal number of cut-off values for different applications. Semivariogram modelling will quickly become tedious if it has to cope with a very large number of cut-off values. Better solutions exist in the realm of stochastic simulation, which allows geographical modelling in the presence of non-stationarity, heterogeneity, and other forms of complexity to a large extent. The simulation paradigm will be covered in the next section, where geostatistics and the broader discipline of spatial statistics offer potential approaches to uncertainty modelling (Diggle *et al.*, 1998).

5.3 STOCHASTIC MODELLING

5.3.1 Analytical Plausibility versus Simulation Versatility

Derivative variables such as slope and aspect may be generated from known variables such as elevation. While various efficient options exist for creating close approximations to derivatives, it is important also to be able to analyse the uncertainty in the outputs given knowledge of uncertainty in the inputs. In general, this means modelling the effects of uncertainties in source data as they propagate through GIS operations, and reporting these effects in connection with the results of GIS analysis.

There are two types of approaches for spatial mapping and uncertainty modelling: analytical and simulation. Consider a function $\Phi(Z)$ of a random variable Z that is expressed as a vector of dimension n, which is equal to the product of the numbers of rows and columns of a raster-structured field, such that locations may be indexed within the range $[1,n]$. In many applications such as terrain slope modelling, $\Phi(Z)$ may be defined on local neighbourhood of much smaller dimensions, for example a 3 by 3 moving window. The expectation of $\Phi(Z)$ is characterised with a probability density function $pdf(z)$, where $\Phi:R^n \rightarrow R$ and $E[|\Phi(Z)|] < \infty$:

$$E[\Phi(Z)] = \int_{R^n} \Phi(z)\mathrm{pdf}(z)dz \qquad (5.12)$$

If one is able to compute the integral in Equation 5.12 analytically or numerically, then the process is said to be an analytical approach for the calculation of the expectation of a spatial function $\Phi(Z)$. Indicator Kriging described previously may be performed to derive analytical solutions for evaluating expectations $E[\Phi(Z)]$.

However, analytical evaluation is seldom feasible with spatial functions, and numerical calculation may not be efficient either for large values of n. Another problem is that the probability density $pdf(z)$ is often known only up to a normalising factor, which is extremely difficult to compute as in the case of a Gibbs distribution with a partition function not computable in closed form, due to the sheer size of the state space in the corresponding model (Besag, 1974; Ripley, 1992). It will be further shown that, for assessing uncertainty in functions of the general form $\Phi(Z)$, the analytical approach is a complex and time-consuming way of arriving at estimates of expectations.

By definition, the variance of a random variable expresses its variability. For a function of a random variable such as $\Phi(Z)$, variance is an extension of Equation 5.12, evaluated from the integral:

$$\text{var}(\Phi(Z)) = \int_{R^n} (\Phi(z) - E[\Phi(Z)])^2 \, \text{pdf}(z) dz \tag{5.13}$$

But for evaluating covariance matrices for vector functions denoted also by $\Phi(Z)$, modification becomes necessary. A common alternative is to apply the law of variance and covariance propagation. In particular, when continuously differentiable arithmetic operations are undertaken, it seems straightforward to input the differential coefficients of model parameters and their standard errors, along with maps of the variables concerned and their prediction errors.

Let JC_Φ be the Jacobian matrix of the function Φ, which is arranged as:

$$JC_\Phi = \partial\Phi_{s1}/\partial Z_{s2} \tag{5.14}$$

where subscripts $s1$ and $s2$ refer to the locations, which are presumably enumerated within $[1,n_1]$ and $[n_1+1,n_1+n_2]$, respectively. With knowledge of the covariance matrix for Z, COV_Z, the application of the law of variance propagation leads to evaluating the covariance for the function $\Phi(Z)$ as:

$$COV_\Phi = JC_\Phi COV_Z JC_\Phi^T \tag{5.15}$$

where the superscript T denotes the matrix transpose.

An analytical evaluation of variance propagation depends on the existence of information about COV_Z. Geostatistics provides a set of techniques for analytical modelling of uncertainties in random variables including elevations and slopes. As mentioned previously, geostatistics is advantageous as a spatial interpolator in the sense that a squared error measure, $\|Z(i,j)-z(i,j)\|^2$, where i,j identifies a pixel, is minimised in Kriging, and the minimised squared errors are then taken as variances at individual locations. Further, when normality is assumed, the distribution of a random variable is fully characterised by its mean and variance. Maps of estimates and variances obtained by Kriging will be able to provide inputs for the analytical modelling of spatial uncertainty, though subject to a degree of approximation.

Consider a variable, say elevation, over a regular grid. Assume that variances and covariances are estimated for any location and its neighbours. Knowledge of the variances and covariances would permit analytical modelling of variances in derivative variables such as slope. Assume that the tangent of slope for grid location i,j, $slope(i,j)$, is estimated via Equation 2.11, which is itself one of many possible methods for the analytical approximation of slope (Florinsky, 1998). Applying the law of variance propagation, it is possible to estimate the variance in $slope(i,j)$, $\sigma^2_{slope(i,j)}$, via:

$$\sigma^2_{slope(i,j)} = \frac{1}{\tan^2(slope(i,j))} \left[\begin{array}{l} \left(\frac{\partial Z}{\partial X}\right)^2 \left(\sigma^2_{Z(i,j+1)} + \sigma^2_{Z(i,j-1)} - 2\text{cov}_{Z(i,j+1;i,j-1)}\right) \\ + \left(\frac{\partial Z}{\partial Y}\right)^2 \left(\sigma^2_{Z(i-1,j)} + \sigma^2_{Z(i+1,j)} - 2\text{cov}_{Z(i-1,j;i+1,j)}\right) \end{array} \right] \tag{5.16}$$

where $\sigma^2_{Z(i,j+1)}$, $\sigma^2_{Z(i,j-1)}$, $\sigma^2_{Z(i-1,j)}$, and $\sigma^2_{Z(i+1,j)}$ represent the variances of Z at locations denoted by $(i,j+1)$, $(i,j-1)$, $(i-1,j)$, and $(i+1,j)$, respectively, while $cov_{z(i,j+1;i,j-1)}$ and $cov_{z(i-1,j;i+1,j)}$ stand for covariances between location pairs $(i,j+1)$ and $(i,j-1)$, and $(i-1,j)$ and $(i+1,j)$, respectively.

First of all, the derivatives of Equation 5.16 must exist to permit the analytical estimation of variance propagation. But terrain aspect may not be differentiable at locations of undetermined gradient, such as sharp ridges. Secondly, the spatially varying variances and covariances in location-specific elevations that appear in Equation 5.16 should be fully evaluated to recover the spatial variance of slope. Apparently, successful quantification of spatial variability and dependence is the key to modelling uncertainties. However, Kriging variances are actually averaged measures when in reality $Z(x)$ can take all possible values, and are dependent on the data configuration, suggesting that Kriging variance has limited use for the analytical modelling of uncertainties.

A further complication is that Equation 5.16 is not simple, involving different coefficients, variances, and covariances. Analytical approaches are often simplified by assuming the absence of correlation among input data layers and across neighbours, especially when complex mathematics are involved and when there is a lack of knowledge of spatial dependence. For instance, Equation 5.16 may be simplified to $2\sigma^2(1-\rho)$ by assuming uniform spatial variances σ^2 and covariances $\rho\sigma^2$, as has been shown already. Therefore, analytical approaches seem geared only to situations where sources of uncertainty are not significantly correlated, and when the geo-processing performed on the data sets is sufficiently simple to permit arithmetic approximation. A stochastic simulation strategy can be shown to be more versatile, and to be better adapted for solving real-world problems of spatial uncertainty.

Stochastic simulation is the term given to the problem of generating n replicas of Z in the form of $\{z^\tau\}$ that in turn serve to produce a series of $\Phi(Z)$ as $\{\Phi(z^\tau)\}$. The expectation of $\Phi(Z)$ is estimated as:

$$E[\Phi(Z)]^* = \frac{1}{n}\sum_{\tau=1}^{n}\Phi(z^\tau) \tag{5.17}$$

where τ cycles from 1 to n, the total number of realisations produced for $\Phi(z^\tau)$. With sufficiently large value of n, one will be getting quite close to the true $\Phi(Z)$ on average. By extension, the covariance for $\Phi(Z)$ may be approximated as:

$$COV^*_{\Phi(Z)} = \frac{1}{n-1}\sum_{\tau=1}^{n}\left(\Phi(z^\tau)-E[\Phi(Z)]^*\right)\left(\Phi(z^\tau)-E[\Phi(Z)]^*\right)^{\mathrm{T}} \tag{5.18}$$

The following two sub-sections discuss in detail two types of methods for stochastic simulation. One is geostatistical, and is based on a useful extension of Kriging, while the other is built upon Markov-chain Monte Carlo methods. The general idea behind stochastic simulation is to use the computer to reproduce numerically what is known of the underlying spatial distributions by means of investigation and knowledge, giving rise to location-specific data and, more importantly, to the characterisation of spatial dependence. The complicated background and lengthy process of stochastic simulation as compared to easier analytical error propagation are balanced by a greater concern for the fundamental principles of digital mapping and analysis.

5.3.2. Geostatistical Approaches

Kriging provides the best linear unbiased estimate from the data, as established previously in Chapter 3. Kriging estimation cannot claim to be any closer to reality than other spatial interpolators solely on the grounds of the spatial dependence incorporated for optimal weighting, as it does not restore completely the spatial variability believed to exist in the underlying phenomenon. For one thing, semi-variances computed from Kriged values appear much more regular than the actual ones. An illustration is provided in Figure 5.4, where the Kriged estimate $z(x)^*$ appears much smoother than the unknown reality of the random variable $Z(x)$; the two functions are shown as the solid and the darker dashed lines, respectively, in profile.

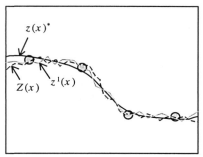

Figure 5.4 $z^1(x)$ and $z(x)^*$ as stochastic simulated and Kriged versions, respectively, of field $Z(x)$.

By contrast, distributions are more realistic if they honour sampled data and reproduce the spatial dependence observed in the data, or deemed applicable in the light of knowledge from similar spatial distributions. An example realisation is shown in Figure 5.4 as $z^1(x)$, and is clearly more typical than the Kriged version $z(x)^*$ of the underlying distribution $Z(x)$.

This is well catered for in geostatistical simulation, also known as stochastic imaging, whereby an uncertainty model is used to generate a set of equal-probable samples conforming to a prescribed distribution (Journel, 1996). The samples generated from a stochastic simulation are often called realisations of the model and provide the basis to drive analysis of uncertainty. Thus, by construction, simulations of alternative realisations of the same random function must have the same characteristics as those revealed by histograms and semivariograms constructed from the data. Simulation approaches have been advocated as a more versatile technique for propagating uncertainties than analytical approaches, as the former can be applied to any situations where methods exist for generating realisations (Goodchild, 1995).

What would be valuable for stochastic simulation with spatial dependence is the conditional distribution of error at location x, given the measurements, say $z(x_s)$, $s=1,...,n$, that are used for estimation at that location. Hence, it is possible to retain the measured values at sample locations to assure coincidence with what is known of the phenomenon. This can be implemented via conditional simulation, as will be described next.

Geostatistical simulation is based on the principle that, by the nature of simple Kriging in the case of stationary variables, estimator and estimation error are independent (Delfiner and Delhomme, 1975; Journel and Huijbregts, 1978). Let a stochastic realisation $z(x)$ of an underlying variable $Z(x)$ be written as:

$$z(x) = z(x)^* + \left(z(x) - z(x)^*\right) \tag{5.19}$$

where $z(x) - z(x)^*$ is the estimation error for an estimator $z(x)^*$, which is usually unknown as $z(x)$ is yet to be determined.

Using the same data set, it is possible to compute Kriging estimates $z_E(x)^*$ and write:

$$z_E(x) = z_E(x)^* + \left(z_E(x) - z_E(x)^*\right) \tag{5.20}$$

The error term $z_E(x) - z_E(x)^*$ becomes known, and can be substituted into Equation 5.19 to derive a new field $z_C(x)$ as:

$$z_C(x) = z(x)^* + \left(z_E(x) - z_E(x)^*\right) \tag{5.21}$$

where $z_C(x)$ has the same structure of spatial dependence as $z(x)$. If a conditional simulation is sought, then it is straightforward to require that Kriging estimates honour the sample values at sample points $x_s (s=1,...,n)$. Thus, it is easy to write $z_E(x)^* = z_E(x)$, $z(x)^* = z(x)$, and finally $z_C(x) = z(x)$.

The simulation process above, which includes both unconditional and conditional forms, may be defined to generate conditional realizations of $Z(x)$ directly, using such methods as sequential Gaussian simulation. Consider the joint distribution of n random variables Z_s with n very large, where the N variables Z_s ($s=1,...,n$) represent the variable Z at the N nodes of a dense grid discretising the field A. By the Gaussian sequential simulation approach, all univariate conditional cumulative distribution functions (ccdf) of Z_s are assumed Gaussian, and conditioned to the original data and all previously simulated values (denoted by |data). Thus, the n-variate ccdf is written as:

$$F(z_1,...,z_N | \text{data}) = \text{prob}\{Z_s \le z_s, s=1,...,n | \text{data}\} \tag{5.22}$$

Successive application of the conditional probability relation shows that drawing an n-variate sample from Equation 5.22 above can be done in n successive steps, each using a univariate ccdf. The procedure for a Gaussian sequential simulation is thus as follows:

1) determine the univariate cdf $F(z)$ for the entire field A;
2) perform the normal score transform of z-data into y-data with a standard normal cdf;
3) define a random path by which each node of the grid is visited;
4) use simple Kriging with the normal score semivariogram model to determine the means and variances of the ccdf of the random variable Y_s for locations s ($s=1,...,n$);
5) draw a value y_1^τ from the univariate ccdf of Y_1 given original data, and update the conditioning data set to a union of the newly simulated data $\{Y_1 = y_1^\tau\}$ with the original data set;
6) proceed to the next node, and loop until all n random variables Y_s are simulated with increasing levels of conditioning;
7) back-transform the simulated normal values $\{y_s^\tau, s=1,...,n\}$ into a simulated joint realisation $\{z_s^\tau, s=1,...,n\}$ of the original variables.

Sequential indicator simulation is widely used, and is more flexible than sequential Guassian simulation in that it is possible to adopt different semivariogram models for different cut-off values, and it also permits incorporation of soft data such as constraint intervals, as seen in Section 5.2. However, it requires the somewhat tedious use of a large number of indicator semivariogram models. Sequential indicator simulation is designed to estimate class proportions and to reproduce indicator semivariogram models; the statistics of the continuous variable being simulated, z-histogram, and semivariogram may be poorly reproduced.

5.3.3 Markovian Stochastic Simulators

A well-known property of Markov random fields is their local dependence in space and time. Let S be a finite set with elements denoted by site s, which has a broader meaning than location. Let Λ be another finite set standing for the phase or state space. A random field on S with phases in Λ is a collection $\{z(s), s \in S\}$ of random variables $Z(s)$ with values in Λ. Thus, a random field can be regarded as a random variable taking values in the configuration space Λ^S. A configuration $z \in \Lambda^S$ is of the form $z=\{z(s), s \in S\}$, where $z(s) \in \Lambda^S$ for all $s \in$ S.

For a given subset SS $\subset$ S, define $z(SS)=\{z(s), s \subset SS\}$. Let S\SS denote the complement of SS in S. It is meaningful to write $z = \{z(SS), z(S\backslash SS)\}$. For a fixed $s \in$ S, S\$s$ is a short form for S\$\{s\}$. A neighbourhood system on S, denoted by N=$\{N_s, s \in S\}$, is a family of subsets of S such that, for all $s \in$ S, N_s contains the neighbours of site s. The couple (S,N) defines a graph or topology, in which S is the set of vertices and N defines the edges. The boundary of SS $\subset$ S is the set ∂SS = $\{\cup_{s \in SS} N_s\}$\SS. Further, N_s^+ denotes the set $N_s \cup \{s\}$.

With the definitions above, a random field Z is called a Markov random field with respect to a neighbourhood system N if for all sites $s \in$ S, the random variables $Z(s)$ and $Z(S\backslash N_s^+)$ are independent given $Z(N_s)$, that is:

$$\text{prob}(Z(s) = z(s) | Z(S \backslash s) = z(S \backslash s)) = \text{prob}(Z(s) = z(s) | Z(N_s) = z(N_s))$$ (5.23)

for all $s \in$ S and $z \in \Lambda^S$. The local characteristic of the Markov random field Z at site s is represented through function π as:

$$\pi_s(z) = \pi(z(s) | z(N_s))$$ (5.24)

the family $\{\pi_s, s \in S\}$ of which is called the local specification of the Markov random field Z.

An important type of Markovian fields is produced by spatially autoregressive processes. In Chapter 2, a spatially autoregressive process was modelled via Equation 2.22. Assume that the expectation of variable Z is moved from Z to the error term ε, which then becomes inflated with a mean of $E[Z]$ that is usually non-zero. Actually, field ε may consist of random or systematic errors, if Z contains trends, discontinuities, and other spatial non-stationarity, as modelled by Arbia *et al.* (1999) in the context of raster data analysis with multi-band images. Equation 2.22 may be re-written with the understanding that the value at site s_1 is influenced only by the values at sites denoted by s_2 within its locality in a linear manner as:

$$z_{s1} = \rho \sum_{s2} w_{s1s2} z_{s2} + \varepsilon_{s1}$$ (5.25)

where ρ is a parameter indirectly determining the strength of spatial correlation,

w_{s1s2} is a weight reflecting the influence of site $s2$ on site $s1$, and ε_{s1} is a noise term, which is often modelled as having a Gaussian error distribution with mean $E(Z)$ and common variance σ^2.

First-order spatially autoregressive processes are commonly modelled by assigning weight w_{s1s2} a value of 1 if $s1$ and $s2$ share a common boundary, and 0 otherwise. Such a spatial autoregressive process is known as the nearest-neighbour model (Bartlett, 1967; Besag, 1974). Illustrations for four- and eight-neighbours with regular tessellation are shown in (a) and (b) of Figure 5.5, respectively, where circles stand for the locations under consideration, and their neighbours are represented by filled circles. More flexibility is possible by using a decreasing function of the distance between $s1$ and $s2$, or an increasing function of the length of common boundary, which is apparently useful for irregular neighbourhoods (Goodchild, 1980).

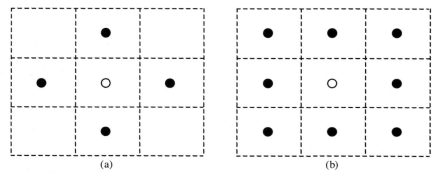

(a)   (b)

Figure 5.5 Regularly tessellated nearest neighbourhoods: (a) four neighbours; and (b) eight neighbours.

Consider random variables z_{s1} ($s1 \in S$), which are often arranged as a row vector Z of length n in a scanning order, where n denotes the number of discretised locations in the field domain. Stochastic simulation is possible based on Equation 5.25 by first obtaining a vector of noise arranged similarly to Z and solving for Z. That is, given error vector ε, vector Z is evaluated from:

$$Z = (I - \rho W)^{-1}\varepsilon \tag{5.26}$$

where I is a unit matrix, and W is the matrix containing elements w_{s1s2}, both being n by n.

However, the matrix solution of Equation 5.26 is computationally intensive. A more efficient solution for field $\{z(s)\}$ is based on iteration, an approach which enabled the modelling of error in categorical data by Goodchild *et al.* (1992) and was later formalised in a Markov framework by Heuvelink (1998). This iterative process is as follows:

1) generate an initial noise (error) field ε from a Gaussian random process of mean $E[Z]$ and variance σ^2 in grid format;
2) test each grid cell in the field against the sum of the values of its four neighbours (the Rook's case—left, right, above, and below—shown in Figure 5.5(a)) multiplied by a supplied value of ρ, which must be in the interval [0,0.25], plus the initial error value at the central grid cell;
3) if the difference between the two quantities is small enough, then the tested

value is deemed to be spatially correlated with its neighbours and output to a temporary data file; otherwise, the original value is replaced with the latter value and recorded in the temporary output file;

4)repeat steps 2 and 3 if necessary to ensure spatial correlation for all grid cells; the result, when fully tested, is a realisation of a spatially correlated field with a particular error distribution.

Note that when the spatial correlation parameter ρ is set to 0 no spatial correlation is assumed, and thus step 1 alone suffices to produce a random error field. But as ρ approaches 0.25, the algorithm takes longer to converge to a solution for spatially correlated Z fields. Upon completion of a field Z, this process may be iterated further but from alternative initial error fields to generate other equal-probable versions of the same underlying field Z.

The issue remains as to the specification of the spatial correlation parameter ρ, which is central to the simulation of a spatially autoregressive process. King and Smith (1988) discussed the connections between geostatistical and spatially autoregressive models via a finite difference equation and other expressions, and Heuvelink (1998) made such approaches accessible to the GIS community. However, simulation may provide a more accessible means for quantifying and driving stochastic processes in space and time, and thus help to broaden the horizon of geographical problem-solving.

The discussion above has implicitly concentrated on the spatial domain. The spatio-temporal context is actually more general and computationally oriented. Spatial simulation amounts to a time process defined over space. Stochastic models generating independent and identically distributed random variables, which behave more or less in the same way, are not always interesting. In order to introduce more variability, one can allow for some dependence on the past. In Markov processes, any probabilistic dependence on the past is only through the previous state, but this limited amount of memory suffices to produce a diversity of behaviours. Discrete-time homogeneous Markov chains possess the required feature, since they can always be represented by a stochastic recurrence equation:

$$Z^{\tau+1} = F\left(Z^{\tau}, Y^{\tau+1}\right) \tag{5.27}$$

where $\{Y^{\tau}\}$ is an independent and identically distributed sequence, independent of the initial state Z^{0}.

Consider a stochastic process $\{Z^{\tau}\}$ that is defined over a finite state space Λ. If for all integers $\tau \geq 0$ and all states $\{z^{\tau}\}$:

$$\text{prob}\left(z^{\tau+1} \middle| z^{\tau}, z^{\tau-1}, ..., z^{0}\right) = \text{prob}\left(z^{\tau+1} \middle| z^{\tau}\right) \tag{5.28}$$

where both sides are well-defined, this stochastic process is called a Markov chain. If, in addition, the right-hand side of Equation 5.28 is independent of τ, this Markov chain becomes homogeneous.

In the following text, elements of states $\{z^{0}, z^{1}, ..., z^{\tau}\}$ are simplified for exposition as the sequence of superscripts $\{0, 1, ..., i, j, ..., \tau\}$. It is convenient to write a matrix $P = \{p_{ij}\}_{i,j \in \Lambda}$, where $p_{ij} = \text{prob}(Z^{j} = z^{j} \mid Z^{i} = z^{i})$, and this matrix is known as the transition matrix of the homogeneous Markov chain $\{Z^{\tau}\}$. Since the elements of the matrix are probabilities, it is required that:

$$p_{ij} \geq 0 \text{ and } \sum_{i \in \Lambda} p_{ij} = 1 \text{ for all states } i, j \in \Lambda \tag{5.29}$$

Denote the probability distribution of the random variable Z at initial state z^{0}

as p_0. Using Bayes's sequential rule and noting the property of homogeneous Markov chains, it is possible to deduce that:

$$\text{prob}\left(z^0,...,z^{\tau-1},z^\tau\right) = \text{prob}\left(z^0\right)\text{prob}\left(z^1\middle|z^0\right)..\text{prob}\left(z^\tau\middle|z^{\tau-1},...,z^0\right)$$

$$= p_0 p_{0,1}...p_{\tau-1,\tau} \tag{5.30}$$

which is valid for all $\tau \geq 0$, thus constituting the probability distribution of the Markov chain $\{Z^\tau\}$.

Further, from Bayes's rule of exclusive and exhaustive causes, it is possible to write:

$$\text{prob}\left(z^j\right) = \sum_{i\in\Lambda}\text{prob}\left(z^i\right)p_{ij} \tag{5.31}$$

which specifies the global balance equation of the form $\pi=\pi P$, and defines a stationary distribution of the transition matrix P, or the corresponding homogeneous Markov chain. Thus, if a Markov chain is started with a stationary distribution, it will keep the same distribution for ever, because then the chain in a state of equilibrium does not depend on the iteration length of τ.

Suppose that an irreducible aperiodic homogeneous Markov chain $\{Z^\tau\}$ is constructed with stationary distribution π over a finite state space Λ. As Λ is finite, the chain is ergodic, and for any initial distribution $\pi(Z^0)$ and all $z\in\Lambda$:

$$\lim_{\tau\to\infty}\text{prob}\left(Z^\tau = z\right) = \pi(z) \tag{5.32}$$

which means that, when τ is large, the distribution approached by Z^τ is close to stationary.

The key to Markov chains lies in the construction of an ergodic transition matrix $P=\{p_{ij}\}_{i,j\in\Lambda}$ on Λ, with a stationary distribution π being the target distribution. Among many such transition matrices, a reversible chain is defined such that:

$$\pi\left(Z^i = z^i\right)p_{ij} = \pi\left(Z^j = z^j\right)p_{ij} \tag{5.33}$$

where elements p_{ij} are evaluated as:

$$p_{ij} = q_{ij}\alpha_{ij} \text{ for } z^j \neq z^i \tag{5.34}$$

where $Q=\{q_{ij}\}_{i,j\in\Lambda}$ is an arbitrary irreducible transition matrix on Λ.

When the present state is z^i, the next tentative state z^j is chosen with probability q_{ij}. When $z^j\neq z^i$, this new state is accepted with probability α_{ij}. Otherwise, the next state remains the same as state z^i. Thus, the probability of moving from z^i to z^j when $z^i\neq z^j$ is p_{ij}, as evaluated by Equation 5.34.

It remains to select the acceptance probability α_{ij}. Hastings (1970) proposed a general form:

$$\alpha_{ij} = q_{ij}^1\middle/\left(1+q_{ij}^2\right) \tag{5.35}$$

where q^1_{ij} stands for elements in a symmetric matrix Q^1 such that $Q^1=\{q^1_{ij}\}_{i,j\in\Lambda}$ and $q^2_{ij}= \pi(Z^i=z^i)q_{ij}= \pi(Z^j =z^j)q_{ji}$.

One can check the reversibility condition in Equation 5.33 which confirms that π is a stationary distribution by a detailed balance test. To satisfy the constraint $\alpha_{ij}\in[0,1]$, the following must hold:

$$q_{ij}^1 \leq 1+\text{minimum}\left(q_{ij}^2,q_{ji}^2\right) \tag{5.36}$$

where equality leads to the Metropolis algorithm (Metropolis *et al.*, 1953):

$$\alpha_{ij} = \text{minimun}\left(1, \pi\left(z = z^j\right)q_{ij} / \pi\left(z = z^i\right)q_{ji}\right) \qquad (5.37)$$

This immediately suggests another line of approach. The sequential Gaussian simulation described previously involves using inverse distributions, where a sequence of random variables embedded in $\{Y^\tau\}$ that corresponds to a normal-score transform of Z data is generated from well-known Gaussian distributions. The random sequence of $\{Z^\tau\}$ actually required is obtained by back-transform. Distribution functions π_Y other than Gaussian may be adopted for such inverse distribution methods. A problem, however, is that π_Y is often known only up to a normalising factor. The desire for simulations of large systems of random variables using such approaches is at the heart of the Markov-chain Monte Carlo method. Indeed, algorithms such as that of Metropolis *et al.* elegantly avoid the problem of standardisation of probability distribution functions, as normalising factors cancel out on the right-hand side of Equation 5.37 (Besag, 1974; Brémaud, 1999).

Several useful stochastic models are based on Gibbs distributions and their historical background of statistical mechanics, which are formally equivalent to Markov random fields (Spitzer, 1971). An important construct for formulating Gibbs distributions is cliques, which consist of specific locations and subsets of more than one element where any two distinct sites are mutual neighbours. On the configuration space Λ^S, real potential functions can be defined describing local interactions relative to subsets of cliques, from which energy functions may further be derived. Gibbs distributions take the form below:

$$\pi_{TP}(z) = \frac{1}{Z_{TP}} \exp\left(-eng(z)/TP\right) \qquad (5.38)$$

where $TP > 0$ is the temperature, $eng(z)$ is the energy of configuration z expressed as a potential function, and Z_{TP} is a normalising factor or partition function.

Applications of Gibbs distributions have been particularly abundant in image analysis, with breakthroughs made by Geman and Geman (1984), who exploited the naturally convenient mechanism of energy functions for describing image properties such as textures and edges. Bayesian image analysis utilising Gibbs prior and posterior distributions has been greeted with increasing interest in spatial statistics. An application is to maximise posterior distributions, as objective functions, given prior models and observed data. In this process, a sequence of images is generated, converging to the maximum *a posteriori* estimation, a type of spatial optimisation. In order to avoid being trapped into local maxima, stochastic simulation permits changes that decrease posterior distributions, which are made on a random basis.

The Gibbs algorithm for spatial optimisation works as follows. Using a strictly positive probability distribution $\{q_s, s \in S\}$, the transition from state $Z^\tau = z^\tau$ to state $Z^{\tau+1} = z^{\tau+1}$ is made by changing the value of the state at one site only with probability $\pi(Z^{\tau+1}(s) \mid Z^\tau(S\backslash s))$ according to its local specification. Site s is chosen independent of the past with probability q_s, which, in Gibbs notation, is $\exp(-d(eng)/TP)$, where $d(eng)$ is change of energy, that is, $eng(Z^{\tau+1})-eng(Z^\tau)$. The new configuration is $Z^{\tau+1} = (Z^{\tau+1}(s), Z^\tau(S\backslash s))$. This gives rise to the non-zero entries of the transition matrix $P = \{q_s\pi(Z^{\tau+1}(s) \mid Z^\tau(S\backslash s))\}$, corresponding to an irreducible and aperiodic chain.

Distributions for q_s are functions of a global control parameter TP, which leads to essential uniformity at high values but, at low values, concentration on states that increase the chosen objective functions. Post-processing of simulated fields takes advantage of the non-parametric solutions offered by simulated annealing, with spatial covariances being taken as objective functions. The gradual reduction of temperatures provides a schedule for simulated annealing, which, when implemented for uncertainty modelling, means that simulated fields can be brought gradually to match conditioning data, and to reproduce the spatial covariance observed in the data. Applications of Gibbs distributions and Bayesian methods are particularly relevant to the spatial statistical analysis of uncertain geographical data, such as DEMs and remotely sensed images.

This section has striven for coherence in the description of stochastic methods that can be used to explore spatially distributed phenomena. The basic aim throughout has been to find ways of usefully mapping spatial uncertainties, through a mixture of analytical and simulation methods, and a range of spatial statistical models. In summary, a wide range of models are available for working with uncertainty in geographical information, and the choice between them depends in part on the application, and in part on the researcher's conceptualisation of the workings of the geographical world and its representation.

5.4 MAPPING UNCERTAIN TERRAINS

5.4.1 Uncertainties in Elevation and Slope

Uncertainties in terrain elevation and slope remain topics of continuing and active research, as has been hinted frequently. This section is intended as a casebook detailing how geostatistics can be applied for modelling the uncertainties occurring in elevation and slope data, with the latter being derived from the former. It will be demonstrated by examples that stochastic simulation is very powerful for furnishing a statistically sound and spatially explicit basis for exploring uncertain terrain-related data.

Suburban-area mapping involves a range of mid-scale data products, usually incorporating base-map data such as buildings, contours, street networks, and land records on the one hand, and environmental data such as soil, water, and noise levels on the other. For this reason, an Edinburgh suburb in the vicinity of Blackford Hill, shown in Figure 5.6, was chosen as the test site to illustrate the modelling methods. Zhang (1996) may be referred to for further detail.

As shown in Figure 5.6, there is a significant amount of variation in topography and thematic entities at the test site, which is dominated by residential buildings, gardens, parklands, roads, and railways. The lower left part of Figure 5.6 shows the slopes of Blackford Hill, a small lake (Scottish loch), worked allotments (small garden plots), and a mixture of vegetated land. Topographic mapping in such an area, with its superimposed physical and cultural entities, is bound to be challenging, raising problems of efficiency and accuracy.

A set of 1:24,000 scale aerial photographs was used to generate test data, while another set of 1:5,000 scale aerial photographs provided the basis for sampling reference data. The 1:24,000 scale aerial photographs in natural colour were flown in mid-June 1988, and it is a subset of this series that is shown in Figure 5.6. The 1:5,000 scale aerial photographs are in natural colour and are part of an experimental sortie of high-resolution material flown for the Ordnance Survey in late July 1990 (Kirby, 1992).

Figure 5.6 An aerial photographic subset showing the Edinburgh suburban test site.

Ground control was established by land surveying, which, in turn, enabled the implementation of a photogrammetric block adjustment using the 1:5,000 scale aerial photographs. This block consists of two strips of photographs, each of five stereo-models, providing photocontrol for photogrammetric digitising based on the 1:5,000 scale aerial photographs. One stereo-model using the 1:24,000 scale aerial photographs was oriented using the set of GCPs obtained jointly by the land surveying and the block adjustment.

Using the 1:5,000 and the 1:24,000 scale aerial photographic stereo pairs, photogrammetric heights were estimated using an analytic plotter to produce reference and test data sets over a smaller area of approximately 400 m by 400 m centred to the north of Blackford Hill. A test was carried out to provide initial estimation of the height measurement accuracy of points digitised from the 1:24,000 scale aerial photographs. In this test, 15 well-defined points, including outstanding landmarks, centres of manholes, and road junctions, were checked against the points obtained from the block adjustment of the 1:5,000 scale aerial photographs. The check points must be withheld from use in controlling the 1:24,000 scale stereo-model, so that independence in checking is not violated. The RMSE in the elevation data is about 0.62 m.

Rather than average uncertainty over space, the spatial variability of elevation and its effect on slope calculation were explored with Kriging and simulation, using the geostatistical software package GSLIB (Deutsch and Journel, 1998). For this, it is necessary to compute experimental semivariograms and then create suitable semivariogram models. Thus, the test data digitised from the

1:24,000 scale aerial photographs were overlaid with the reference data acquired from the 1:5,000 scale aerial photographs. Then, it was possible to compute errors in elevation as the differences between the reference and the test elevation values at individual data points. Test elevation values were interpolated linearly, using a TIN structure. This process resulted in an extended ARC/INFO point attribute data (.PAT) file comprising IDs, reference coordinates, and errors in elevation for 175 points. Maximum absolute differences were restricted to less than 3 m to avoid outliers. Data were reorganised from the .PAT file to build a new data file suitable for GSLIB, in order to calculate semivariograms for elevations, and errors in elevations at individual data points. Also calculated was the spatial covariance for errors in elevations. Results are shown in Figure 5.7.

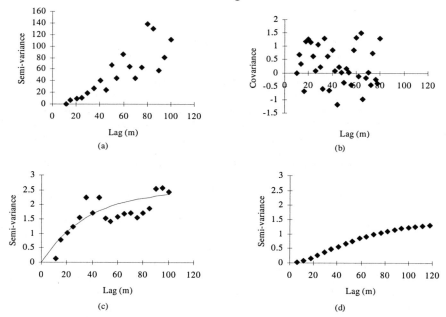

Figure 5.7 Modelling spatial dependence: (a) semivariogram of elevation data; (b) covariance of elevation error; (c) semivariogram of elevation error; and (d) semivariogram of Kriged error.

As shown in Figure 5.7(a), the experimental semivariogram for elevation data behaves like a power function. The spatial covariance for errors in elevations is shown in Figure 5.7(b), and seems randomly distributed. The experimental semivariogram for errors in elevations reveals a typical semivariogram structure, shown as diamonds in Figure 5.7(c), and is fitted with a spherical model semivariogram (the solid line in Figure 5.7(c)).

The results above suggest that the semivariogram for errors in elevations should be used in subsequent Kriging and simulation. Kriging was carried out, with specified parameters including the grid cell size (4.16 m by 4.16 m), together with the range, sill, and nugget effects describing the spherical semivariogram model shown in Figure 5.7(c). Geostatistical simulation was then performed using a Gaussian sequential simulation program SGSIM provided in GSLIB. This began with a normal score transformation and semivariogram modelling. Details are shown in Figure 5.8, where experimental data, an exponential model, and the chosen spherical model are represented by diamonds, dashed lines, and solid lines.

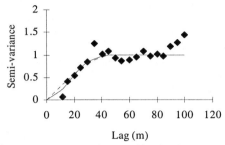

Figure 5.8 Normal score transform and semivariogram modelling.

A set of 30 simulated surfaces of elevation errors was generated, and nine versions are shown in Figure 5.9. As expected, they appear varied yet correlated. A reference surface of elevation was created from a TIN-based interpolation of spot heights sampled from the 1:5,000 scale aerial photographs. Thirty simulated surfaces of errors in elevations were then subtracted from the reference surface to get simulated samples of elevation data with error distributions comparable to that of the original test elevation data set sampled from the 1:24,000 scale aerial photographs.

A comparison between Kriging and conditional simulation is provided in Figure 5.10, where Kriged errors, Kriged elevation, and Kriging variance are depicted in grey-scale surfaces in Figures 5.10(a), (c), and (e), respectively. Figures 5.10(b), (d), and (f) are based on further simulation studies to be described below.

The 30 simulated surfaces of elevation error were summarised by means and standard deviations, as shown in Figure 5.10(b) and (f), while an elevation surface based on the mean of simulated error surfaces is presented in Figure 5.10(d). The surface of standard deviation of errors, shown in Figure 5.10(f), does not register zeros for grid cells containing data points, that is, spot heights. This is due to the fact that, while spot heights were, as usual, assumed to be measurements at points of zero dimension, the grid cells that overlap but do not coincide with data points were artefacts of the rasterisation of the error data.

An important issue concerns the effects of Kriging on the semivariograms of resultant fields. Take the Kriged error in elevation data, shown in Figure 5.10(a) as an example. Its semivariogram is shown in Figure 5.7(d). Clearly, the Kriged error surface appears smoother than the semivariogram model in Figure 5.7(c) would suggest.

At this point, it is interesting to examine the distinction between a simulated error surface and a mean simulated error surface in the light of semivariograms. The results shown in Figure 5.11 confirm that a simulated surface is drawn from the population of possible realisations of the underlying surface being modelled, but the mean of such surfaces is not a possible sample of the population, contrary to a fundamental statistical theorem for non-spatial data (Englund, 1993).

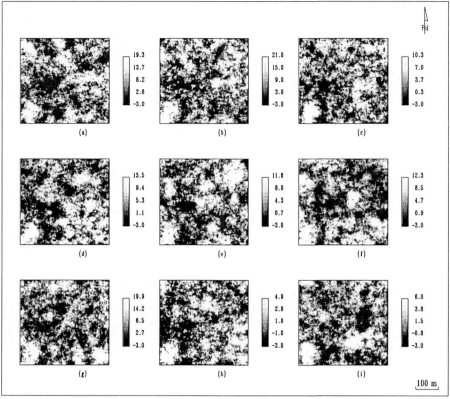

Figure 5.9 A set of nine surfaces of elevation errors shown in (a)–(i).

For modelling uncertainty in slopes derived from elevation data, both simulation and analytical approaches were tested. For simulation-based studies, 30 versions of slope data were calculated from the 30 versions of simulated elevation data. Nine of these slope surfaces are shown in Figure 5.12, where versions (a)–(i) correspond, respectively, to the simulated elevation data containing errors shown in Figures 5.9(a)–(i).

For analytical modelling, on the other hand, the elevation surface interpolated using Kriging along with the surface of Kriging standard deviation shown in Figures 5.10(a), (c), and (e) were used to calculate slope and to estimate error propagation. Statistical analysis of simulated versus Kriged data sets for error propagation in slope estimation generated the graphical displays in Figure 5.13.

Surfaces of estimated slope and standard deviation are shown in (a) and (b) of Figure 5.13 for the Kriged data set, and in (c) and (d) for the simulated data set, while (e) and (f) represent those resulting from the mean of simulated elevation data. Examining Figure 5.13 suggests that there is an apparent difference in the standard deviations in slopes derived from analytical and simulation approaches. Remarkably, the surface of standard deviations estimated from Kriged data, shown in Figure 5.13(b), seems to resemble the typical appearance of the surface of Kriging standard errors shown in Figure 5.10(e). A similar link can be drawn between the surfaces of standard deviations estimated based on the mean of the simulated data set, i.e., the surfaces shown in Figures 5.13(f) and 5.10(f).

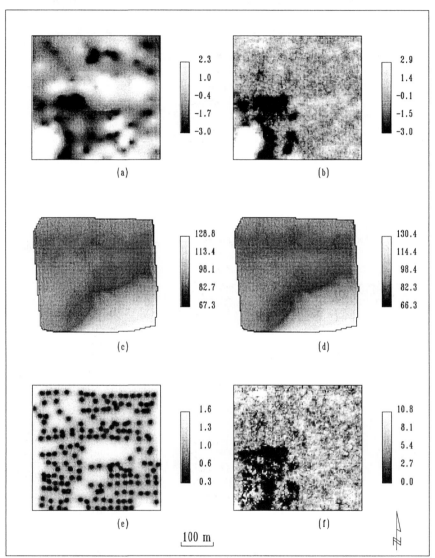

Figure 5.10 DEMs and errors: (a), (c), (e): Kriged error, elevation, and standard error; (b), (d), and (f): mean of simulated error, elevation, and standard deviation respectively.

A tabular summary of the results obtained for uncertainty in terrain modelling by analytical photogrammetry is provided in Table 5.1. On the errors in elevation data, as anticipated, the means of errors tend to be estimated quite uniformly by Kriging, realisations, and mean simulated data. The spatial variability of a simulated elevation error surface is greater than that produced by Kriging. As averaging tends to suppress variations, the mean standard deviation of simulated errors is reduced by half. In other words, standardised Kriging variance exceeds that estimated by averaging simulated error data, as the former is measured by anticipating all possible estimates and thus contains more variability than the latter.

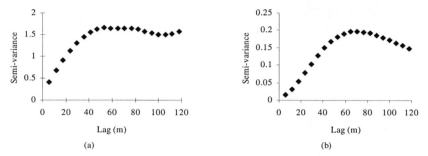

Figure 5.11 Semivariograms for elevation error surfaces: (a) simulated; and (b) mean of simulations.

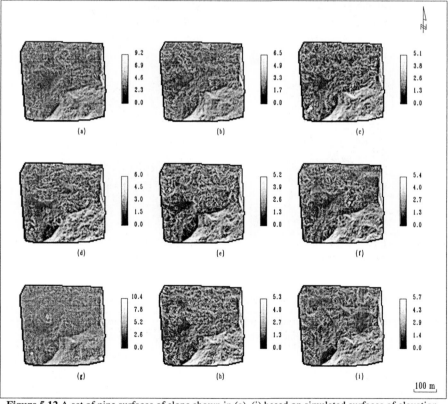

Figure 5.12 A set of nine surfaces of slope shown in (a)–(i) based on simulated surfaces of elevation.

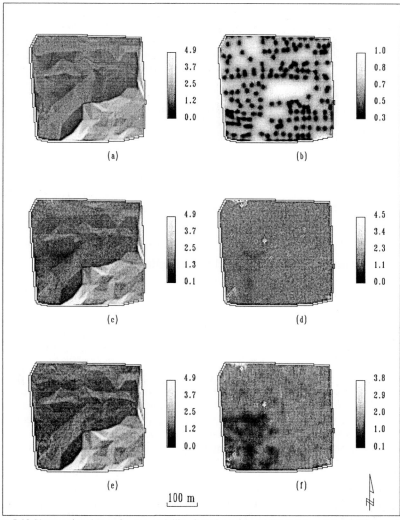

Figure 5.13 Slope estimation and error modelling based on: (a) and (b) Kriging; (c) and (d) the set of 30 simulated elevation surfaces; (e) and (f) the mean of simulated elevation data.

Table 5.1 Modelling errors in elevation and slope (std. = standard deviation).

Elevation error (m)						Slope and its error propagation					
Kriging		Simulation				Analytical				Simulation	
		realisation		mean		Kriged		simulated			
mean	std.	mean	std.	mean	std.	mean	std.	mean	std.	mean	std.
-1.2	1.0	-1.3	1.4	-1.2	0.7	0.75	0.65	0.79	0.84	1.00	0.40

For slope estimation and propagation of errors in elevation data, on the other hand, the story becomes more dramatic. Simulation tends to overestimate slope but underestimate error propagation, as shown in Table 5.1. The reason for slope overestimation by simulation is that simulation generates more varied elevation surfaces, which imply more chances for slope estimation to become inflated. Another possibility is that simulated data may be contaminated with computational artefacts, further increasing data variability. But, as elevation data errors are not purely random but spatially correlated, the derivative-based formula for slope calculation shown in Equation 2.11 has the effect of cancelling out a certain proportion of the effects due to elevation data errors.

The use of analytical error propagation for slope estimation suggests that averaging simulated data leads to estimates of slope that are closer to those produced by Kriged data, as shown in Table 5.1. However, error propagation estimated by the analytical method exceeds that based on Kriged data. An apparent explanation is that simulated elevation data are more spatially varied than the Kriged data.

5.4.2 Uncertainties in DEMs Created from Digital Photogrammetry

The foregoing subsection has examined errors and their propagation with respect to digital elevation and slope data, based on spot heights sampled from an analytical plotter. In that process, elevation data were supposedly sampled from the ground, subject to machine limitations and human errors such as stereoscopic acuity. Such manual techniques for topographic mapping are heavily dependent on trained and fatigue-resistant photogrammetrists and traditional mapping techniques, although certain of the processing steps may be computer-enhanced.

Digital photogrammetry is able to produce DEMs consisting of densely and economically sampled elevation data semi-automatically. However, if mapping is performed in a wooded or urbanised terrain, digital photogrammetric techniques may result in elevation data of fine resolution but inordinately inferior accuracy due to gross errors in image matching (Baillard and Dissard, 2000). Ambiguity exists in mapping terrain containing vegetated surfaces or other artefacts if these overshadow more natural terrain features, since the computer can seldom match the interpretative intelligence of humans, and hence stereoscopic acuity. Similar problems have been identified with respect to airborne laser scanner data that could potentially shed light on photogrammetric data (Schreier *et al.* 1984; Kraus and Pfeifer, 1998). Figure 5.14 shows a situation where photogrammetric measurements $z_2(x)$ comprise a thicker profile that may contain data points whose elevations are affected by treetops and buildings, as are often indicated by the abruptness of a profile.

Co-Kriging may be thought of as an obvious option for deriving estimates of terrain surfaces. As indicated in Figure 5.14, due to the existence of non-terrain segments in photogrammetric data, the geostatistical combination of sparse spot heights and dense photogrammetric data tends to produce surfaces that may stay close to terrain where that terrain is well represented by both data sources, but can deviate significantly in other areas. A more flexible alternative is the simulated annealing approach, by which an optimal but compromised surface is obtained by maximising a chosen set of objective functions, including the probability distribution of the terrain surface as sampled by spot heights and conditional to digital photogrammetric data. The capability of simulated annealing for estimation of terrain on the basis of multiple data sources relies on the proper encoding of prior distribution information and spatial variability. However, the imposition of non-terrain segments in digital photogrammetric data is likely to deal a severe blow to the non-parametric elegance inherent in simulated annealing.

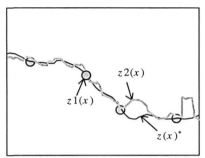

Figure 5.14 Discerning errors in elevation data $z2(x)$ created from digital photogrammetry with the aid of spot heights $z1(x)$, and used to estimate terrain $z(x)^*$.

How digital photogrammetric data can be processed to remove the uncertainty imposed by disturbance and the intrusion of entities foreign to genuine terrain surfaces provides the key for operational topographic mapping. Kraus and Pfeifer (1998) tested a method of airborne laser-scanner data filtering and terrain-surface interpolation. It works iteratively such that an initial surface is computed using equal weights for all height measurements, but is later adjusted with weights that depend on the initial residuals. As vegetated locations are more likely to have positive residuals, a weighting function can be defined so that real terrain measurements are weighted with 1, and non-terrain measurements with 0, while points lying in between receive weights that are inversely proportional to the residuals.

A geostatistical model for mixed populations was proposed by Zhu and Journel (1991), which amounted to the probabilistic mapping of the spatial potentials of parent populations prior to the weighted averaging of measurements from respective populations. This requires information about population identities of sampled locations, which, in the case of terrain modelling, is dependent on the correct interpretation of terrain and non-terrain locations in the first place. This section adopts an approach that makes use of residual surfaces computed by testing a set of digital photogrammetric data against reference spot heights, thus enabling the differentiation of terrain and non-terrain surface segments by appropriate thresholding of residuals.

Successful data analysis should provide improved estimates of the real terrain based on relatively sparse sampling of spot heights and dense photogrammetric measurements. Terrain modelled this way should be closer to the true terrain than using either data set alone. An illustration is given in Figure 5.14, where terrain is estimated as $z(x)^*$ from collaborative use of accurate spot heights $z_1(x)$ and photogrammetric data $z_2(x)$, with the latter subject to lower accuracy and the unknown superimposition of non-terrain entities.

A test was carried out at the site shown in Figure 5.6. Reference spot heights were sampled from the 1:24,000 scale photographic stereo pair on the same analytical plotter employed for the tests reported in the previous subsection. This set of reference spot heights, 357 in total, was divided into two subsets, one for interpolating terrain surfaces (257 points), and the other for validation (100 points), as shown as triangles and squares in Figure 5.15(a), respectively.

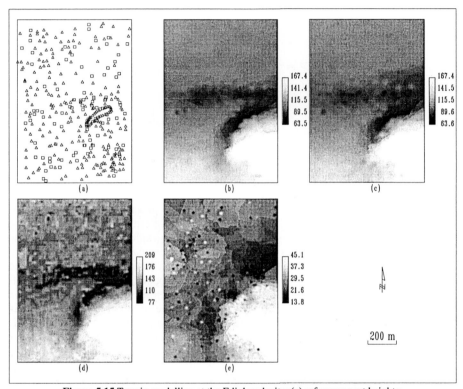

Figure 5.15 Terrain modelling at the Edinburgh site: (a) reference spot heights; (b) reference elevation surface; (c) elevation surface using a subset of the reference spot heights; (d) digital surface data produced from digital photogrammetry; and (e) error surface.

The totality of spot height data was used to produce the reference surface shown in Figure 5.15(b), while the subset of 257 spot heights generated the surface shown in Figure 5.15(c). Surfaces shown in Figures 5.15(b) and (c) were interpolated using the Kriging procedures provided by GSLIB (Deutsch and Journel, 1998). Experimental semivariograms and their models are displayed as diamonds and lines, respectively, for the whole set of 357 spot heights in Figure 5.16(a) and the subset of 257 spot heights in Figure 5.16(b). Both sets of data were fitted with Gaussian models.

Using a desktop scanner to scan the 1:24,000 scale aerial photographs, a digital image pair with pixel size of about 4.16 m at ground resolution was loaded onto a PC running a desktop mapping software system. Inner and exterior orientations were accomplished with sub-pixel accuracy, based on a few measured points from the analytical plotter.

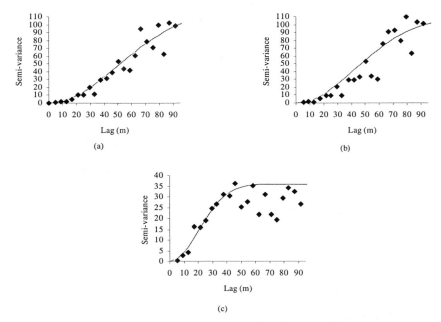

Figure 5.16 Semivariograms: (a) 357 spot heights; (b) 257 spot heights; and (c) elevation error data.

Image matching performed with a target window of 7 by 7 pixels led to the production of the layer of DEM data shown in Figure 5.15(d). This set of digital photogrammetric DEM data had an extent of 286 rows by 193 columns and a square lattice spacing of 4.16 m, and covered a photographically visible surface that included both terrain and other physical and cultural entities. The DEM data set shown in Figure 5.15(d) enabled rectification of the model-area image into the digital ortho-image at ground resolution of 4.16 m shown as Figure 5.6.

The errors or mismatches between the 257 spot heights obtained by digital photogrammetry and their reference heights were found to lie between 17.3 m and 42.7 m. The error data were then calculated for variogram modelling in Gaussian form, as shown in Figure 5.16(c). This allowed for Kriging of elevation errors, the result of which is shown in Figure 5.15(e). The error surface of Figure 5.15(e) appears rather irregular and indicates significant systematic errors.

Co-Kriging the spot heights and digital photogrammetric data directly gave rise to the surface shown in Figure 5.17(a), where segments of non-terrain entities such as buildings, shadows, and trees are discernible. As hinted previously, simulated annealing may be usefully pursued for obtaining estimates of terrain by combining both data sources. A number of subjective decisions have to be made in simulated annealing. These include designing annealing schedules, and weighting of competing objective functions, whose optimisation forms the goal of the

computationally intensive process. Simulated annealing generated 50 realised
surfaces utilising spot heights and digital photogrammetric data, from which
surfaces of mean elevation, lower 0.05 quantile, and upper 0.95 quantile elevation
were derived, as shown in Figure 5.17(b), (c), and (d), respectively. It is possible
to state that the true elevation values at individual locations lie between the lower
and upper *p*-quantile surfaces with a probability of 90%. Figures 5.17(b)–(d)
indicate that surfaces produced by simulated annealing are more dominated by
elevation value ranges set by the spot height data than by those generated by co-
Kriging, and also appear smoother. An interpretation is that simulated annealing
tends to produce surfaces combining different data sources, and thus compromises
the differences more than co-Kriging.

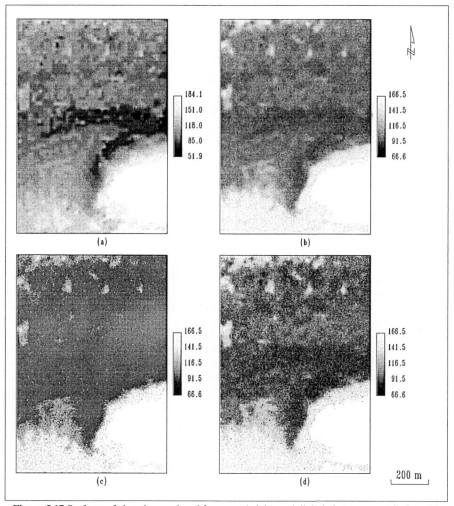

Figure 5.17 Surfaces of elevation produced from spot heights and digital photogrammetric data: (a)
mean by co-Kriging; (b) mean; (c) 0.05 quantile; and (d) 0.95 quantile by simulated annealing.

The surfaces shown in Figure 5.17 were estimated by compromising spot height data, and the possible surfaces they imply, with the surface sampled at finite resolution by digital photogrammetry. If digital photogrammetric data contain measurements pertaining to both terrain and non-terrain entities, the uninformed combination of spot heights and such digital data is sure to end up with neither genuine terrain nor true landscapes, but something reconciling their discrepancies. These discrepancies are potentially interesting, but they should not be confused with measurement errors in digital photogrammetry. As implied by Equation 3.29, inaccuracy in digital image matching has significant impacts on accuracy in the resulting height data. Visualising errors in elevation data in Figure 5.15(e) suggests a severe level of overestimation by digital photogrammetry that would be unreasonable even for rudimentary reconnaissance, let alone for operational topographic mapping.

Further data analysis was necessary to explore the possibility of filtering out errors in digital photogrammetric data and of achieving increased accuracy of topographic mapping by the combined use of reference spot heights and digital photogrammetric data of high resolution. For this, the DEM data set created by digital photogrammetry and shown in Figure 5.15(d) were corrected for the error surface shown in Figure 5.15(e), resulting in the corrected version of the digital photogrammetric grid shown in Figure 5.18(a).

The corrected DEM data set shown in Figure 5.18(a) was then compared with the DEM surface shown in Figure 5.15(c), which was Kriged using the set of 257 reference spot heights, to generate a residual surface. This surface of residuals is shown in Figure 5.18(b), and seems to have sharpened the distinction between terrain and non-terrain surfaces. A probability-like surface was mapped from the residual surface via an exponential function, $\exp(-|\text{residual}|/4.0)$, which was used for the discrimination of terrain from non-terrain. This probabilistic map is shown in Figure 5.18(c), and should be easier for visual interpretation.

The range of residuals shown in Figure 5.18(b) is between −43 m and 26 m, and trial and error hinted at a threshold of 2 m for discerning the authentic terrain surface from foreign materials superimposed on the terrain. Thus, 18,945 grid cells were identified as being part of the terrain surface, and Kriging was performed for the corrected digital photogrammetric data using only the 18,945 grid cells identified above. The Kriged surface is shown in Figure 5.18(d), and appears closer to the real terrain than its predecessor shown in Figure 5.15(d).

Co-Kriging was performed to tap further the potential of spot height data and digital photogrammetric data. The semivariogram for the set of 257 reference spot heights was already quantified and modelled, as shown in Figure 5.16(b). Co-Kriging using the reference spot heights and the Kriged digital photogrammetric data shown in Figure 5.18(d) was undertaken to produce the further adjusted surface shown in Figure 5.18(e).

The processing detailed above capitalised on mismatch data recording the differences between the digital photogrammetric data and the reference data, thus enabling the identification of terrain and non-terrain segments in the digital photogrammetric surface data. To gain further understanding of the advantages of co-Kriging based on error removal, as opposed to the direct use of digital photogrammetric data for combined estimation, an accuracy assessment was employed. Accuracy was evaluated with respect to the mean and standard deviation of errors by testing all the 357 reference spot heights against the surface generated using Kriging. These were calculated for the DEMs produced by Kriging with the 257 reference spot heights, co-Kriging and simulated annealing using spot heights and digital photogrammetric data, Kriging using the corrected digital photogrammetric data approximating the real terrain, and co-Kriging using the spot heights and processed digital photogrammetric data. Results are reported in Table 5.2. The accuracy for the final stage of DEM data processing described above was also evaluated using the 100 validation spot heights shown as squares in Figure 5.15(a), and the results are also listed in Table 5.2.

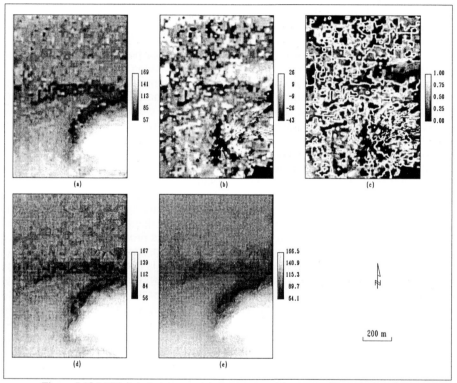

Figure 5.18 (a) Digital photogrammetric data corrected for errors; (b) residual surface; (c) discerning terrain segments from digital photogrammetric data; (d) Kriged terrain surface based on (c); and (e) co-Kriged terrain surface based on 257 reference spot heights and the data shown in (d).

As is seen in Table 5.2, the DEM derived with spot height data alone has standard deviations in errors of 2.6 m and 4.1 m over the whole surface and at validation locations, respectively. For DEMs created from the combined use of spot heights and digital photogrammetric data, standard deviations in errors rise to twice those levels, and by half over those obtained using spot heights alone for surface-wide and location-specific assessment, respectively. Although the approach of simulated annealing produced slightly better results than those from co-Kriging, the large errors reported with the combination of spot heights and digital photogrammetric data reveal that computational models may be very effective but are no substitute for accurate data. If the process of differentiating between terrain and non-terrain segments is performed, the resulting errors are close to those based on spot heights alone. While surface-wide evaluation suggests error levels close to those obtained using spot heights alone, location-specific assessment indicates that errors drop for the DEM created from Kriging terrain-approximating segments of digital photogrammetric data, but rise closer to the error levels measured for the DEM created using spot heights alone. Furthermore, it is interesting to note that there are only modest differences in mean DEMs created by the different methods, except for those based on the use of spot heights and the original digital photogrammetric data.

Table 5.2 Accuracy evaluation of DEMs derived from geostatistical processing of analytical and digital photogrammetric data (unit: metres).

Methods for data processing	Surface-wide		Location-based	
	mean	std.	mean	std.
Kriging spot heights	-0.1	2.6	0.1	4.1
Co-Kriging spot heights and digital photogrammetric data	-1.2	6.9	-1.0	7.0
Simulated annealing using spot heights and digital photogrammetric data	-0.5	6.0	0.1	6.1
Kriging digital photogrammetric data approximating terrain	0.2	2.9	0.1	3.3
Co-Kriging spot heights and terrain-approximating digital photogrammetric data	0.0	2.6	0.0	3.8

The general interpretation is that suburban fabrics behave somehow irregularly and unpredictably in producing terrain measurements by digital photogrammetry that are sharply different from those that result from an assumed model derived from a few spot heights that were identified visually. One should not blame digital photogrammetry alone for the large errors observed, as these errors may occur due to differences in the entities being measured, as illustrated by Figure 5.14. These methods raise basic issues concerning what is being tested, and which source provides the reference for the other in terms of deviation. Errors in high-resolution elevation data, such as those created from digital photogrammetry, may be validated against reference spot heights, but such an evaluation is implicitly applicable to these specific spot heights only. If the reference data are spatially extended to form surfaces, and used to drive a surface-wide accuracy assessment of high-resolution data, the results will contain measurement errors and other differences due to the models adopted for terrain surfaces. For instance, the latter differences in elevation may be caused by non-terrain entities such as ditches, trees, and buildings, which are filtered out in conceptual terrain models but captured by remote sensors.

Digital photogrammetry must also be assessed from the perspectives of speed and repeatability, against which it is believed to be superior compared to visual and manual height estimation; the latter by contrast is costly, and dependent on training and skills, and subject to human fatigue. As the previous test has confirmed, inaccuracy in digital photogrammetric mapping is a consequence of errors in image matching. With quality images and robust image matching, digital photogrammetry can be superior to quality elevation data from image sources. But terrain modelling based on sparsely sampled data of supposedly higher accuracy has yet to be combined effectively with densely measured data of lower accuracy, such as those acquired from digital photogrammetry, so that suburban complexity can be mapped more accurately at reduced cost.

In order to demonstrate the promising aspects of digital photogrammetry in terms of measurement accuracy, some further tests were conducted. Digital photogrammetry was carried out again but based on a university neighbourhood at the south foot of LuoJia Hill in central China's Wuhan City, where Wuhan University is centred. A sophisticated digital photogrammetric workstation running the VirtuoZo software system was employed; this software system was produced by a strong team of academicians and researchers at the university, and has been widely adopted in China and abroad (Zhang *et al.*, 2000). An aerial photographic pair flown on 6 October 1993 at a scale of 1:3,000 was scanned using a quality scanner set at 25 microns, resulting in digitised images each of 9,040 rows by 9,040 columns. An extract of the aerial photograph showing the Wuhan test site is depicted in Figure 5.19(a).

A group of undergraduate photogrammetrists performed ground surveys of horizontal and vertical control networks using a combination of theodolite-based angular and distance observations, and levelling of triangulation stations and benchmarks. Four GCPs were used for reconstituting a digital stereo-model covering the Wuhan test site, with two validation points reporting residuals of less than 0.2 m after completion of interior and exterior orientations.

Following stereoscopic reconstitution of the digital image pair, image matching was undertaken in an endeavour to automate the production of DEMs. As was stated in Chapter 3, the reliability of image matching may be strengthened by correlating pixels along epipolar lines rather than in a small rectangular window, where conjugate pixels are assumed to be found. The epipolar line-based strategy was adopted by the VirtuoZo developers, resulting in improved efficiency in image matching. A grid DEM was generated from digital matching of the digitised image pair, and set at a ground resolution of 0.5 m; this DEM is shown in Figure 5.19(b). The ortho-image shown in Figure 5.19(a) was produced at a ground resolution of 0.2 m to match the mapping scale of 1:1,000.

The digital photogrammetric workstation provided users with by-product maps showing the reliability of image matching based on correlation coefficients and other image-matching criteria. The reliability surface for image matching is shown in Figure 5.19(c), and is scaled between 0 and 100. The harp-shaped lake was found to be home to some relatively unreliably correlated image segments, thus calling for caution in regard to the elevation data derived therein.

An empirical assessment of the accuracy of digital photogrammetric height estimation was carried out by means of 83 surveyed spot heights, whose distribution is shown in Figure 5.19(d). Gross errors of mean value 22.54 m were revealed, which occurred due to mismatching, blurred texture and, above all, shadows and loss of image due to the viewing angle of the aerial photography. This set of erroneous points is marked with tiny squares over the grid shown in Figure 5.19(b). The removal of gross errors led to a mean of –0.2 m and an RMSE of 0.54 m in height, while setting a much tighter error threshold of 1 m gave rise to a mean of -0.15 m and an RMSE of 0.40 m.

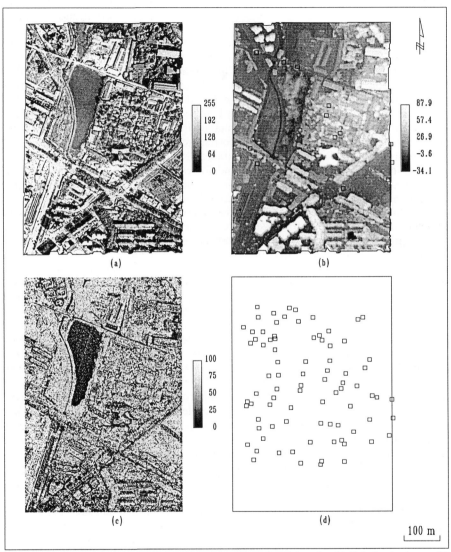

Figure 5.19 Terrain modelling by digital photogrammetry at the Wuhan site: (a) digital ortho-image; (b) DEM; (c) reliability of image matching; and (d) validation data.

It is confirmed that, at finer image resolution and with an improved solution for image matching, digital photogrammetry was able to produce DEMs of reasonable accuracy. Also notable is the fact that computer vision has yet to reach the level of intelligence needed in geographical measurement, because current forms of computerisation of topographic mapping serve to augment but are not yet able to replace humans, who still are able to attain a higher standard of accuracy. In this respect, feature extraction is still not fully supported by digital methods.

5.5 DISCUSSION

This chapter has examined how errors in continuous variables can be described and modelled using a variety of techniques. As uncertainty descriptors, surfaces of standard errors may be derived as by-products of Kriging, or more objectively given the availability of validation data. Errors can be treated as special types of fields, which provide crucial information about the underlying fields being mapped in terms of their variability. More informative are cumulative distributions of variables at unsampled locations, conditional to sampled data, which can be obtained from indicator Kriging and used for estimating confidence intervals for unknown locations. Data products, such as mean fields, may be derived from interpolated vector fields of cumulative distributions for the underlying fields.

As has been argued by many authors, uncertainties are not simply matters of accuracy assessment by testing against reference data of higher accuracy, but rely on the quantification and incorporation of spatial dependence for objective evaluation of errors as they find their way into information products. In order to overcome the burden of modelling spatial covariances for a large number of cut-off values in indicator Kriging and, more importantly, to combat real-world complexity, stochastic simulation has been described as a better option than analytical approaches for modelling errors as they propagate through geographical information processing. In doing so, a coherent introduction has been given to analytical and simulation-based approaches to tracking the propagation of errors in source data. Stochastic modelling is seen to be central to much spatial statistical analysis of data errors and, ultimately, to the uncertain geographical distributions being investigated.

With real data sets covering a suburb and a university community, geostatitics has been explored for describing elevation data errors and modelling their propagation in slope estimation. The integrity of spatial variation has been seen to be preserved via quantification, and through the reproduction of spatial dependence in elevation and slope. Without the mechanisms of stochastic simulation, uncertain geographical distributions may remain elusive and impose unpredictable consequences on information processing.

Kriging and stochastic simulation provide effective visualisation of uncertainty. While error surfaces can be output to emphasise one view of spatially varying errors, confidence intervals were depicted to highlight a spatial extension of descriptive statistics, as has been seen in the experiment with simulated annealing via the combined use of spot height and digital photogrammetric data. Stochastic simulation generates equal-probable realisations of a field, which are effective metaphors for reinforcing concepts of uncertainty in geographical phenomena, in which data are honoured and spatial covariance is duplicated. Such concepts lie at the core of geographical inquiry.

As documented in Section 5.4.1, the estimation and modelling of error surfaces stand as one possibility for mapping uncertainty. In this process, it is a prerequisite that reference data are available to measure differences in the data set being assessed. Implicitly assumed therein is that the assessed data and the reference data are co-located, so that it is meaningful to perform location-specific validation of a particular data set. This is rather demanding and hard to satisfy in real-world scenarios where mapping and validation surveys may be subject to spatial and temporal gaps, and where sampling may not cover the full geographical extents of field variables. Thus, for many applications concerning accuracy assessment for continuous variables, errors are estimated by interpolation using either the data set under investigation or the reference data, or both. A side effect is that some spatial variability may be filtered due to interpolation, while other new errors are introduced. Extra caution should be exercised in mapping and modelling error surfaces. However, another straightforward approach is by direct stochastic simulation of the variables under study. This will circumvent the requirement for complete availability and co-location of the data being assessed and the reference data. If location-specific errors are to be assessed, they can be obtained by overlaying a realised data surface and the reference data points or vice versa. Error surfaces retrieved in this way should be closer to the spatial variability and dependence deemed appropriate for the variables under study.

Heterogeneous data are increasingly brought into GIS for spatial problem-solving. However, there is a lack of interpretability and compatibility between disparate sources of data, which may include, on the one hand, ground-based data that are believed to be accurate and, on the other hand, remotely sensed data of high resolution but inferior accuracy. The empirical tests reported in the previous section have shown the practicality of terrain modelling, where various semantic differences and computational difficulties have hindered the co-operative use of multi-source data. Turning such data diversity into information wealth remains a topic for further extensive research.

On the topic of digital photogrammetric mapping, two issues remain prominent. One is related to the possibility for enhancing the accuracy of digital photogrammetric measurement at the level of individual locations. The other is more fundamental, and relies on the effectiveness and adaptability of computer techniques for differentiating terrain and non-terrain from high-resolution data before using them in improved topographic mapping and, broadly, geographical problem-solving.

The separation of terrain and non-terrain measurements by digital photogrammetry is not only beneficial for terrain mapping, but also for mapping other collateral variables such as vegetation, and discrete objects such as buildings and roads. For example, image correlation for photogrammetric mapping in forested regions requires shadow removal and ground visibility in images to track the terrain surface precisely, while accurately measured surfaces containing forested tracts can, in principle, allow for the mapping of forest qualities and timber quantities given knowledge of terrain surfaces. A technique that employs a series of templates of typical species to identify trees was explored by Quackenbush *et al.* (2000) with promising results. As terrain surfaces must reside below surfaces measured from image models, such knowledge may be encoded by indicator Kriging to drive the optimal estimation of terrain surfaces.

The quest for improved terrain modelling in operational settings suggests the adaptation of geostatistical approaches appropriate to the modelling of complex terrain. As terrain becomes heterogeneous and non-stationary, flexibility and robustness of data handling over uncertain terrain are of particular significance. For this, additional spatial statistical methods such as Markov chains and Gibbs samplers exist for sensible adaptation and implementation. In particular, simulated annealing may be used for terrain and land mapping in general by combining data from visual and computer-based estimation. Image intensity data, along with height measurements from reference and digital photogrammetric sources, may be subjected to Bayesian analysis facilitated by Markovian local specification and Markov-chain Monte Carlo techniques.

These developments in stochastic simulation are of very significant value to geographical information scientists, and to broader communities interested in working with geographical information. While geostatistical approaches have been the mainstream techniques implemented for exploring spatial uncertainties with real terrain data, stochastic paradigms in general will have far-reaching impacts on geographical information processing. As computational techniques are increasingly adapted for and put to geographical use, stochastic methods will become more widely accepted within geographical information communities. The strategies followed and methodologies pursued in this chapter will be useful not only for continuous variables but also for categorical variables and discrete objects, as later chapters demonstrate. It is anticipated that, with the advancement of information technologies, the computerised handling of geographical uncertainties will be less subject to misuse, and less dominated by digits of spurious precision. Only then will it be possible to extract a full range of geographical insights from the growing abundance of available geographical information.

CHAPTER 6

Uncertainty in Categorical Variables

6.1 INTRODUCTION

Many spatially distributed phenomena are necessarily abstracted into discrete-valued categories (a term used interchangeably with classes in this book), and so are the outcomes of spatial discretisation, for the purposes of management, evaluation, and decision-making. A commonly known geographical example for categories is land cover, which may be differentiated into such nominal types as grassland, shrub, and woodland.

Remote sensing is applied to an increasing extent for mapping natural resources and environments. Categorical data are an important output of remote sensing when it is applied in geographical studies. However, misclassifications, which can be interpreted as errors in classified data, continue to hinder the efficient application of remote sensing and its integration with GIS. A variety of reasons can be cited, including data limitations such as the presence of pixels of mixed class (mixels) due to the sensor's finite and coarse spatial resolution; ill-designed or ill-defined classification schemes; the inadequacy of algorithms to handle noisy data; the inability to discriminate classes from one another in respect to certain features; and, above all, the complexity of the real world. The net result is that most classified remotely sensed images have only modest accuracy in operational terms. Moreover, it is not easy to improve accuracy, despite rapid advances in the technology and algorithms of remote sensing and image processing (Townshend and Justice, 1981).

The problem of inaccurate classification is, unfortunately, not limited to digital classification, since the skill of humans at interpretation is far from perfect. For land cover mapping using aerial photographic interpretation, for example, the analyst has to allocate heterogeneous land patches to prescribed categories, which are largely discrete, based on his or her interpretative and subjective judgment. Because of the highly varied and limited contrast between some land cover types in terms of tones and textures, it is not uncommon for photo-interpreted data to be of limited accuracy and consistency, even when high-quality aerial photographs are used (Edwards and Lowell, 1996).

To facilitate the description and modelling of uncertainty in categorical data, spatial categories are viewed as categorical variables that are single-valued functions of location, in other words fields (Goodchild *et al.*, 1992). In that sense they complement the fields of continuous variables such as terrain elevation and slope that were discussed in the previous chapter. They also form a major source of data for GIS-based spatial modelling and applications. This chapter will consider uncertainty in categorical variables. Because there are many more facets to the concepts of randomness and vagueness in the case of categorical variables, the issue of uncertainty tends to be more complicated than in the case of continuous variables, in spite of the simple appearance of categorical maps.

To develop a systematic strategy for discussing uncertainties in categorical variables, it will be necessary to examine the conceptual and methodological issues underlying categorical mapping. Conventionally, categorical variables are depicted in the form of exhaustive, non-overlapping areal units separated by a network of boundary lines, where single discrete labels are ascribed to individual areal units, on the assumption of internal homogeneity. This representational model has been termed the area-class maps (Mark and Csillag, 1989) or the categorical map (Chrisman, 1989), and is often displayed in the form of mosaics of coloured categories delineated as polygons, or contiguously grouped raster cells.

Certain well-defined and self-explanatory categories, such as water bodies, are particularly suitable for discrete representation, usually because of strong correlation between the associated categories in the conceptual (information) and observational (data) domains. Such categories lend themselves to efficient classification by remote sensing techniques with reasonable accuracy, posing no major problems of uncertainty as far as categorisation is concerned. Such categories may be mapped with inaccurate boundaries, as is the case with remotely sensed images of coarse spatial resolution, but the categories themselves are nevertheless distinct and well-defined.

Discretised categories may also result from inventory and managerial requirements. In agriculture and forestry, for example, spatial categories such as agricultural plots and forest stands are necessarily conceptualised as discrete units and usually depicted as irregular polygons. In such cases any within-class heterogeneities are suppressed or discarded, and the mapped boundaries between classes reflect such exogenous phenomena as historical field boundaries and the boundaries of clearcut forest stands, and are consequently well-defined. Similar examples can be found in the processes responsible for the definition of property boundaries or the boundaries of statistical reporting zones and administrative units, which are rarely related to the properties or attributes reported for the units themselves.

Classification accuracy in discrete categorical data is usually assessed by sampling and constructing error matrices using available reference data. A useful measure of classification accuracy is the percentage of correctly classified locations, which may also be reported at the level of individual classes. Such classification accuracies are commonly interpreted as probabilities, and averaged over space or aggregated over categories.

However, many of the spatial categories used to characterise the variation of physical and human phenomena are characterised by spatial heterogeneity. The conceptualisation of complex spatial categories as discrete object-like features may be very misleading, since it is a potential source of inaccuracy. Moreover, because of such complexity in the distribution of categories, many geographical phenomena do not lend themselves to straightforward categorisation by computer, and there are various forms of inconsistency in human observation. The combined effects of various types of non-homogeneity mean that it is hard to argue that each location or areal unit on a map has been classified with equal accuracy, or that a certain class has been mapped with an accuracy uniform to that class. Therefore, categorical uncertainty must be conceptualised as spatially varying, and such a view is consistent with spatially distributed occurrences of classes and the way they are perceived and observed by humans, and by their sensory extensions such as remote sensors. In other words, categorical uncertainty cannot be separated from, but rather must be modelled along with, the underlying categories themselves, and with their spatial variation.

Some heterogeneities in categorical distributions, such as small inclusions that fall below the minimum mapping unit area, or below the coarse spatial resolution of remote sensors, are sensibly considered noise and should be smoothed in practice. Others are better conceptualised as an essential part of the spatial variation of the phenomenon that should specifically be described. Thus, it is often constructive to retain information-rich categorical heterogeneities, rather than to filter them as noise in the interests of simplicity (Robinove, 1981; Lewis, 1998). Such heterogeneity would tend to be suppressed by an arbitrary positioning of boundaries, and provides an argument in favour of the use of fine raster representations rather than irregular polygons.

At first sight, the use of irregular polygon representations may permit more accuracy in the positioning of complex class boundaries. But this would be at odds with mapping categorical complexity, because the process of locating naturally occurring spatial categories is actually secondary to the process of classification. Thus, the need to position boundaries may not enhance or expedite classification, but may instead introduce a substantial element of unjustifiable subjectivity. Conceptually, it may be easier for two people to agree on the class to be assigned at a specific point, than to agree on where the boundary should lie between two classes. It is thus often sensible and worthwhile to use regular areal units for categorical mapping, avoiding the overhead of locating boundaries during classification. In this way, meaningful and spatially coherent categories, in the form of blocks of contiguous grid cells, may be created through the appropriate grouping and splitting of individually classified locations, by exploiting spatial adjacency and other contextual properties. Moreover, remote sensing has reinforced the popularity of raster cells for categorical mapping, by making it easy to acquire images in raster format. Besides, there is an often unspoken minimum area size for meaningful spatial classification, implying that spatial categories are context-dependent and should not be seen as defined over arbitrarily small points. Thus, the adoption of regular areal units is not likely to cause any serious degradation of accuracy for categorical data, particularly since accuracy is not judged merely by geometry.

The adoption of truly field-based representational models facilitates the spatially explicit representation and description of uncertainties in class allocation. If uncertainties in categorical data can be expressed in probabilistic terms, then it is possible to construct stochastic formalisms for uncertainty modelling, and to exploit spatial dependence in combining multi-source data (Goodchild, 1994). Sensitivity to categorical homogeneity or heterogeneity is a prerequisite for the effective use of categorical data, as this chapter will demonstrate.

Uncertainties will have differing and mixed characteristics depending on the techniques employed for categorical mapping. Any spatial classification is likely to be influenced by both objective physical factors and a certain amount of human-related subjectivity. The selection of samples for training and testing a classifier is clearly in part subjective, and may even be somewhat arbitrary. On the other hand, multispectral reflectance data, which are used to drive image classification, contain noise-contaminated yet objective measurements of radiant influx from the segments of terrain within the sensor's instantaneous field of view. While objectivity may suggest a greater degree of tractability in the description and analysis of uncertainty using frequentist concepts of probability, it is also tempting to treat subjectivity using suitable probabilistic methods (Dempster, 1967; Yager, 1991). In reality, both objective and subjective components of uncertainty are likely to be juxtaposed in spatial classification (Johnston, 1968), and probabilistic protocols alone may not always be sufficient for effective modelling.

The subjective dimensions of categorical uncertainty are more important when categorisation and evaluation involve inherently vague concepts, or when the focus is on the artefacts of discrete quantification of categorical continua. For example, a classification of vegetation into categories of healthiness is often vague and may not be as straightforward as it appears at first sight. Similarly, a categorisation of vegetation abundance is frequently highly subjective. The labelling of a type of landform as alluvial sand has an element of vagueness, even though the class definition includes apparently objective properties such as slope, if the threshold for such properties is described using such vague terms as gentle. The geomorphological classification of landforms into moderate- to high-relief areas dominated by slope processes is also inherently vague, although quantitative measurements of slope gradients are obtainable. Urbanisation as a spatial concept suggests a continuum of vague categories ranging from rural and suburban to urban. Accordingly, it makes sense to think of the spatial extents associated with the conceptual continuum as similarly vague and continuously varying from non-membership to full membership, rather than as changing suddenly. Another relevant example is in land suitability analysis, where assessment is commonly based on scoring alternatives against multiple criteria chosen *a priori*, giving rise to graded suitability ratings from extremely unsuitable to highly suitable.

Fuzzy set theory and fuzzy methods, developed in the computer and information sciences (Zadeh, 1965), have been welcomed by many researchers in the GIS and science communities because of their capability for dealing with a continuum of membership grades of inherently fuzzy spatial categories, as reviewed by Fisher (2000). Fuzzy approaches to classification are the subject of continuing research, and are widely accepted as providing more flexible information on categorical continua (Foody, 1996; Townshend, 2000). Fuzzy approaches to evaluating classification accuracy have also been explored using fuzzy error matrices and linguistic quantifiers, and fuzzy categorical data, such as estimates of area, may be presented in terms of levels of fuzzy class memberships (Binaghi *et al.*, 1999; Woodcock and Gopal, 2000).

While fuzzy sets are widely seen as being able to represent inherently fuzzy concepts and linguistic fuzziness, critics are wary of certain negative aspects of fuzzy sets, notably the arbitrary definition of membership functions and, in addition, their relatively short development history and weak mathematical basis. If fuzzy memberships are assigned arbitrarily, then the use of multiple data sources and data reduction methods leads to exponential growth in the number of possibilities of class memberships. Clearly, the probabilistic framework is better grounded mathematically than the younger fuzzy school. The continuous characterisation of class memberships is common to both probabilistic and fuzzy approaches to classification. This kind of similarity should not cause any confusion about the crispness or vagueness of the underlying information classes, as probability and fuzzy sets are both oriented to coping with aspects of uncertainty. Much human thinking and decision-making is vague out of necessity, and solving realistic spatial problems is likely to involve both chance components and elements of vagueness (Zadeh, 1997). De Bruin (2000b) gave examples utilising fuzzily quantified probabilities for spatial queries such as site selection. In light of the co-existence of probabilistic and fuzzy methods, it is not only constructive to see probability and fuzzy sets as complementary bodies of theory, but also important to put them both to operational use.

An issue that is often overlooked concerns the distinction between information and data-driven classes. Rough sets may be of help in understanding this point. Rough sets were defined by Pawlak (1982) to handle uncertainties due to a lack of information about some elements of the universe of discourse. Vagueness arises due to the indiscernibility of certain elements in the context of available data or knowledge. In other words, vagueness results in boundary-line elements that cannot be linked to a particular concept (set or class) or to its complement. This is directly related to the issue of discrepancies between information and data classes, a problem identified also by the remote sensing community in relation to the gap between information and spectral classes. The main advantage of rough sets derives from the claim that they do not need any preliminary or additional assumptions about data, and knowledge is elicited from the data available, unlike basic probability assignment and fuzzy membership characterisation. This chapter explores the use of rough sets as a way of improving the compatibility between information and data classes, with the aim of enhancing the discernibility of otherwise indiscernible classes. By examining various bodies of theory relevant to uncertainty analysis, the chapter highlights some of the distinctive and collaborative aspects of probabilistic, fuzzy, and rough methods for mapping spatial categories such as land cover.

6.2 PROBABILISTIC DESCRIPTION OF SPATIAL CATEGORIES

6.2.1 Probabilistic Fields and the Geostatistics of Categories

Consider a field over the domain D, where an observation or sample $u(x)$ represents the categorisation or classification of location x to one of c mutually exclusive and exhaustive classes $\{U_1,...,U_c\}$. For instance, U_1 may be grassland, U_2 woodland and so on, for classification of land cover. The order of classes in the set may be arbitrary, while for those categories that are the result of transformation from the continuous variables discussed in the preceding chapter, a natural order exists such that the categories are arranged as a set of intervals corresponding to ascending cut-off values of a continuous variable.

A conventional, yet still useful, model for a realised categorical variable such as $u(x)$ is to represent areas of homogeneous class as polygons. The boundaries are represented cartographically by precisely defined lines of zero width, although it is frequently misleading and inaccurate to represent adjacent areal units on categorical maps in this way (Mark and Csillag, 1989). Shown in Figure 6.1(a) is a polygonal map depicting four classes $\{U_1,...,U_4\}$, where location x, being within the area labelled U_1, is believed to belong to class U_1. An evaluation of classification accuracy is required for the reliable use of categorical data. Assume it is possible to obtain independent reference data. Then it is straightforward to cross-tabulate the classified against the reference data, enabling various accuracy measures of the classification to be calculated.

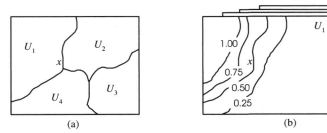

Figure 6.1 Categorical variables shown as: (a) polygons; and (b) probability vector fields.

Using comparisons between classified and reference data, one is able to evaluate the classification in terms of per cent of correctly classified locations, leading to an estimation of accuracy in terms of frequentist probability. This conventional approach to accuracy assessment suggests that either each location is classified correctly into its nominal class with the overall classification accuracy reported or, at a more detailed level, a particularly classified location has a probability of being correctly classified as that particular class. Understandably, certain locations return higher probabilities of being correctly classified than others. As shown in Figure 6.1(a), the proximity of location x to the boundary is obvious. This may make potential map users more hesitant over the reliability of its categorisation as U_1 than locations further away from the boundary, simply by reckoning on the increased heterogeneity and hence likelihood of misclassification near borderlines. Thus, it makes sense to progress from a global or class-based accuracy assessment of class allocation to a location-specific evaluation of probabilities of class occurrence. A similar approach was adopted by Steele *et al.* (1998), who experimented with a spatially based concept of misclassification probabilities. The framework supporting spatially varying class probabilities is better built upon raster structures than polygons in order to accommodate spatial heterogeneity, continuity, and dependence in class occurrence, though clearly the size chosen for the raster cells will affect the results.

It is generally not known beforehand which of the c classes is present at a location. Thus, at any location $x \in D$, a random variable $U(x)$ is defined, taking any of the values from the set $\{U_1,...,U_c\}$. Given a particular set of knowledge and observational data concerning domain D, it is possible to assign discrete classes to individual locations, leading to a classified or realised field of $U(x)$ denoted by $u(x)$, with locator x referring to the domain D, that is, $x \in D$.

For a probabilistic extension of the random categorical field $U(x)$, define the occurrence probability $p_k(x) = \text{prob}(U(x) \in U_k)$, for individual classes $k=1,...,c$. It is useful to construct a probability vector field $\{p_1(x),...,p_c(x) \mid x \in D\}$, whose elements are evaluated as class occurrence probabilities for location x. This formula defines a set of probabilistic fields for variable $U(x)$ pertaining to c categories, as shown in Figure 6.1(b), where a possible specification of the values of the probability vector at location x may be $\{0.55,0.35,0.10,0.00\}$.

To estimate probabilistic fields for a particular set of categorical variables, various approaches have been researched. Suppose c mutually exclusive classes can be found over a domain and n classified samples are available. Each observation is classified as a member of one of the possible classes $\{U_1,...,U_c\}$. An important formalism for categorical data lies in the transformation of the classes

$u(x)$ at the sample location x into binary indicators, a transformation that is precluded in the case of continuous data:

$$i(U_k;x) = \begin{cases} 1 \text{ if } u(x) \in U_k \\ 0 \text{ if } u(x) \notin U_k \end{cases}, k = 1,...,c \tag{6.1}$$

The indicator transform $i(U_k;x_s)$ of $u(x_s)$ may be seen as the realisation of the indicator random variable $I(U_k;x_s)$ at location x_s. For each class U_k ($k=1,...,c$) under consideration, these samples are indicator transformed (i.e., an observation is transformed to 1 if it is classified as class U_k, 0 otherwise), leading to indicator data denoted as $i(U_k;x_s)$, $s=1,2,...,n$, for a particular class U_k.

To proceed with a meaningful discussion of probabilistic categorical mapping, a categorical map is seen as a realisation of a random process possessing per-class probabilities $p_k=\text{prob}(U(x) \in U_k \mid x \in D)$, for $k=1,...,c$. Define the following stationary and isotropic joint probabilities:

$$p_{kl}(dis_{s1,s2}) = \text{prob}(U(x_{s1}) \in U_k, U(x_{s2}) \in U_l \mid x_{s1}, x_{s2} \in D) \tag{6.2}$$

for $k,l = 1,...,c$ and $s1, s2 = 1,...,n$

where $dis_{s1,s2} = |x_{s1}-x_{s2}|$ is the distance between locations x_{s1} and x_{s2}. A nearest-neighbour rule for deriving $I(U_k;x_0)^*$, an estimated realisation of $I(U_k;x_0)$ at an unobserved location x_0, is as simple as $I(U_k;x_0)^*=i(U_k;x_{s1})$, where $dis_{0,s1}$ is the minimum distance from x_0 to a data point. By referring to Equation 6.2, an evaluation of classification accuracy may be obtained from conditional probability:

$$\text{prob}(I(U_k;x_0)=1 \mid I(U_k;x_{s1})=1) = p_{kk}(dis_{0,s1})/p_k \tag{6.3}$$

for $k = 1,...,c$ and $s1 = 1,...,n$

For Equation 6.3 to be useful practically, a bivariate probability function of the form $p_{kl}(d)$, as in Equation 6.2, must be estimated for a continuous range of d values. This was done by Switzer (1975) using gridded samples (with integer multiples of the unit grid spacings as d values). The main limitation, as for any nearest-neighbour rule, is that only a single nearest sample is relied upon for probability evaluation as well as for classification. Although it is possible to consider an n nearest-neighbour rule, it would be difficult to make optimal estimates with multivariate probability functions of the form $\text{prob}(I(U_k;x_0)=1 \mid I(U_k;x_s), s=1,...,n)$, since these depend on the configuration of the available samples.

Geostatistical methods, indicator Kriging in particular, have been explored and increasingly adopted for mapping uncertain categorical data. For a categorical variable, the expectation of its indicator transform leads to the direct evaluation of the conditional probability that a certain class k prevails at location x as:

$$E[I(U_k;x)] = 1 \times \text{prob}(u(x) \in U_k) + 0 \times \text{prob}(u(x) \notin U_k)$$
$$= \text{prob}(U(x) \in U_k)$$
$$= p_k(x) \tag{6.4}$$

for $k = 1,...,c$

The indicator transforms of sample points at x_{s1} and x_{s2} are denoted by $i(U_k;x_{s1})$ and $i(U_k;x_{s2})$, where U_k stands for the class label ($k=1,2,...,c$). The experimental semivariogram can be calculated from the sample data using Equation 5.3 and substituting $i(U_k;x_{s1})$ and $i(U_k;x_{s2})$ for $i(z_k;x_{s1})$ and $i(z_k;x_{s2})$, respectively.

Suppose that an experimental semivariogram has been calculated by using Equation 5.3, and has been subsequently fitted to a suitable model. Indicator Kriging performs the estimation of the indicator transform, that is, the probabilities of finding individual classes U_k ($k=1,...,c$), at a point x using simple indicator Kriging, from the estimating equations:

$$p_k(x) = i(U_k;x)^* = p_k + \sum_{s=1}^{n} \lambda_s \left(i(U_k;x_s) - p_k \right) \qquad (6.5)$$

where $i(U_k;x_s)$ represents the indicator transform of a sample point x_s ($s=1,2,...,n$) to class U_k, λ_s is the weight associated with the sample point x_s, and p_k is the marginal probability of class U_k inferred from sample data (Journel, 1983; Solow, 1986).

When a similar process is applied to every location (cell) on a pre-defined grid, a surface of probability is generated. As there are c classes, each class is often dealt with separately. In other words, for each class, a set of binary data is transformed from classified samples. A suitable semivariogram model is fitted to the corresponding experimental semivariogram, calculated using Equation 5.3. In the end, c probability surfaces are generated for c classes, where the values at a location represent the probabilities of finding candidate classes at that location.

The probability scores as calculated above may not always sum to 1.0 across all candidate classes for each location. In the case of categorical probabilities, the constraint is easily met by checking if the estimated probability values are within the interval [0.0,1.0], and setting those outside the interval to the nearest bound, 0.0 or 1.0.

Indicator Kriging performs the estimation of a conditional probability distribution without making assumptions about the form of the prior distribution functions. When a categorical variable is concerned, such as land cover or soil type, indicator Kriging directly estimates the probabilities of finding individual classes at an unsampled location, given a set of classified samples. This is a remarkable feature for indicator Kriging, which Bierkens and Burrough (1993) demonstrated in their work on probabilistic soil mapping. Indicator Kriging is therefore seen as being superior to other methods, such as the inverse-distance-weighted interpolation method, as confirmed by Zhang and Kirby (1997).

It may be argued that, once a specific set of categories has been described adequately by semivariogram models, there are no major difficulties with respect to their probabilistic representation and analysis. However, questions do remain over the formulation of practical strategies for probabilistic mapping using human interpretation and remote sensing techniques. It is not clear by what criteria classified samples may be identified and whether there is appropriate guidance on sampling spatial categories. The next section provides some details.

6.2.2 Probabilistic Classification of Graphical and Digital Images

The probabilistic indicator formalism developed so far views the occurrence of spatial categories as conforming to certain stochastic and spatially correlated distributions quantified via indicator semivariograms, which are usually inferred from sampled locations of known classes or reference data. Thus, the utility of indicator Kriging for categorical data appears to rely heavily on the process by which classified samples are defined and selected. In principle, it is important that the classified samples be acquired with sufficient certainty. But the subjectivity in specifying certainty leaves categorical sampling with a lingering degree of vagueness, and makes its implementation less straightforward than the theoretical elegance of indicator Kriging might have suggested. It is thus constructive to review some of the mainstream methods for categorical mapping, notably, photo-interpretation and digital image classification, in light of the probabilistic formalism set forth in the preceding section.

Consider categorical mapping from aerial photographs using visual interpretation and manual delineation, for the purpose of data inventory, or for the provision of surrogate ground data. It is widely admitted that an interpretative process is prone to misclassification due to subjectivity and inconsistency, the existence of confusing or unfamiliar categories, and other factors. Given that a photo-interpreter often feels relatively at ease in picking locations that display physical homogeneity as well as visual similarity to categorical prototypes, it seems reasonable to start from these core locations, which are assumed or known to be representative of the nominal classes, and to develop categorical probabilities thereafter. Thus, a reasonable approach is to locate samples in the interiors of relatively homogeneous categories, far away from the boundaries and thus avoiding likely transitional or mixture zones. Referring to Figure 6.1(a), classified samples may be chosen in the manner illustrated in Figure 6.2(a), where squares and circles (solid and open) represent samples of four different classes. With classified samples available, semivariograms are estimated and modelled for indicator-transformed data, and indicator Kriging can then be used for estimating the probabilities of finding candidate classes under consideration at location x.

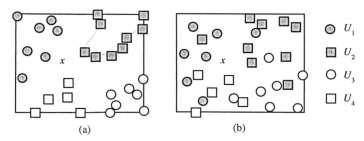

Figure 6.2 Sampling a categorical variable at core locations for estimating classes occurrence at location x: (a) spatially homogeneous; and (b) spatially heterogeneous.

The existence of certain spatially homogeneous classes, such as water bodies, suggests the appropriateness of accommodating deterministic elements within a stochastic framework. This can be accomplished by locating classified samples along the outmost borders of each individual class's core area. An illustration is given in Figure 6.2(a), where the outmost border of Class U_2, shown as solid squares, is marked as grey-scaled lines, and classified samples are located along the border line. Such a sampling scheme leads to the interpolation of probabilistic class occurrence at location x based on TIN-structured indicator sample data, a process that is attractive in retaining the border geometry of discrete class units and general core areas (Zhang and Kirby, 1997). Lowell (1995) provides a useful adaptation of Voronoi diagrams (the dual of TINs) for probabilistic mapping in photo-interpreted forest classes. It is worth noting that indicator Kriging may still be used for situations involving homogeneous categorical segments delimited by discrete lines, as Jupp *et al.* (1988) showed in an attempt to model the discrete distributions of tree canopies.

However, Figure 6.2(a) may result from over-discretisation of originally heterogeneous distributions of categories. It is more objective to retain categorical heterogeneity than to discard it. This is shown in Figure 6.2(b), where spatial heterogeneity is realistically illustrated as inclusions in the otherwise homogeneous categories shown in Figure 6.2(a). While mapping well-studied and broadly defined categories, including simple land-cover types, is generally a manageable task with photo-interpretation, many other natural categories are complex, and photo-interpretation is consequently tedious and vulnerable to the vagaries of subjectivity and bias. Besides, it is often hard to generalise from the rules and experience applied during sampling. Indeed, interpretative sampling of spatial categories is distinct from measurement of continuous quantities such as terrain elevation, where errors are more objectively quantifiable, as has been seen in Chapter 5.

Remotely sensed digital data are known to hold considerable potential for quantifying categorical complexity, and for enhancing objectivity in categorical mapping. They possess the advantages of synoptic coverage, uniform sampling, and flexible configurations of spatial, spectral, and temporal resolution. In addition, through the collateral use of remote sensing data and spatial models inferred from ground data or their surrogates, it is possible to sample and map heterogeneous categories effectively, thus improving classification accuracy and fidelity to reality.

Conventionally classified digital spectral data are, however, of limited use, as has been discussed in Chapter 3, since the date often end with per-pixel allocation of discrete class labels, and assessment of classification accuracy. But it is possible to modify conventional classifiers so that most intermediate results are exploited as sources of valuable information about categorical uncertainty. There are a number of ways to derive probabilistic membership values, depending on the specific classification techniques being used. Assuming normality of spectral (reflectance) data, maximum likelihood classification is based on an estimated probability density function as expressed by:

$$\mathrm{prob}\big(z(x)|k\big) = (2\pi)^{-nb/2} \big|\det(COV_k)\big|^{-0.5} \exp\big(Mdis_{xk}^2/2\big) \tag{6.6}$$

where $\mathrm{prob}(z(x)|k)$ is the probability density function of the spectral measurement $z(x)$ for a pixel x as a member of class k, nb is the number of spectral bands, COV_k is the variance-covariance matrix of class k in spectral space, det denotes the determinant, and $Mdis_{xk}^2$ is the squared Mahalanobis distance from pixel x to the centroid M_k of class k in the spectral space, which is calculated as:

$$Mdis_{xk}^2 = \big(z(x) - M_k\big)^{\mathrm{T}} COV_k^{-1}\big(z(x) - M_k\big) \tag{6.7}$$

where the symbols ^{T} and $^{-1}$ indicate the transpose and inverse of matrices, respectively (Duda and Hart, 1973; Strahler, 1980).

A class label is usually derived by assessing the *a posteriori* probability of membership on the assumption that the pixels belong to one of the pre-defined classes. The *a posteriori* probability of a pixel x with spectral measurement $z(x)$ belonging to class k, $\mathrm{prob}(k|z(x))$, may be determined from the Bayesian rule:

$$\text{prob}(k|z(x)) = p_k \text{prob}(z(x)|k) \sum_{j=1}^{c} p_j \text{prob}(z(x)|j), k = 1,...,c \tag{6.8}$$

where c is the total number of classes, p_j is the *a priori* probability of class j ($j=1,...,c$), and $\text{prob}(z(x)|j)$ is as defined in Equation 6.6.

The measure as calculated by Equation 6.8 in combination with Equation 6.6 provides direct estimation of the elements constituting a probability vector for pixel x, and $\text{prob}(k|z(x))$ can be considered as approximating $p_k(x)$, and denoted as such. Further, assuming equal *a priori* probabilities p_k of classes under consideration, or when no such *a priori* information exists, it is easy to see that $p_k(x)$ can be approximated by normalising $\text{prob}(z(x)|k)$ to unity. Useful work has been done utilising such probabilistic approaches in the remote sensing of spatial categories (Foody *et al.*, 1992).

The probabilistic classification described above is in sharp contrast to a conventional classification, as the latter is concerned with the assignment of discrete labels rather than continuous-valued class memberships. Another type of probabilistic measure may be derived by referring spectral distance values such as the squared Mahalanobis distance ($Mdis_{xk}^2$) to the χ^2 distribution with degrees of freedom equal to the number of spectral bands comprising the reflectance data. This kind of probability is understood as the proportion of pixels belonging to class k with squared Mahalonbis distances greater than $Mdis_{xk}^2$ from the mean of class k.

Statistical approaches, especially those that assume Gaussian distributions of multispectral data, are often found to be of insufficient flexibility and unacceptable accuracy for classifying remote sensing images, because of unrealistic assumptions. To minimise reliance on assumptions about data distributions, non-parametric approaches are especially welcome. Class memberships may be estimated from frequency distributions or histograms obtained for each class from spectral data. In a supervised classification, the class histograms may be constructed from the training data. Such an extended use of histograms constitutes a simple implementation of non-parametric classification, and is well worth exploring.

Artificial neural networks have gained increasing popularity recently in the digital classification of remote sensor images, particularly as they do not rely on distribution assumptions and are able to integrate ancillary data of both continuous and discrete scales. When a neural network is used for classification, continuous-valued activation levels of the network output units corresponding to the set of categories under consideration may be used to derive probabilistic class memberships (Foody, 1996).

Subjective probability affects class membership definition through the evaluation of *a prior* probabilities in Equation 6.8, in addition to various forms of subjectivity entering the process of categorical mapping. The evidence theory approach to classification, which offers explicit mechanisms for encoding subjective beliefs in terms of given evidence, may be adapted for probabilistic categorical mapping (Srinivasan and Richards, 1990). Evidence theory was first formulated by Demspter (1967), and was later enhanced by Shafer (1986) in the context of the probabilistic combination of evidence. In the evidential approach to probabilistic classification, an important concept is the basic probability assignment. This expresses the degree to which all available and relevant evidence supports the claim that a particular element (e.g., a pixel) of the universal set belongs to a particular set (e.g., a land-cover class). Those subsets of the universal set on which the available evidence focuses are called focal elements of the basic probability assignment, and equate to the classes being mapped. A set of focal elements and their associated basic assignments comprise a body of evidence.

Given a basic probability assignment *pm* for a focal element, that is, a category label *k*, its belief measure *Bel*(*k*) and plausibility measure *Pl*(*k*) can be determined from:

$$Bel(U_1) = \sum_{U_0 \subseteq U_i} pm(U_0) \tag{6.9}$$

and:

$$Pl(U_1) = \sum_{U_0 \cap U_1 \neq \varnothing} pm(U_0) \tag{6.10}$$

where *Bel*(U_1) represents the total evidence that an element belongs to class U_1 as well as to the various subsets of U_1, and *Pl* (U_1) represents not only the equivalent of *Bel*(U_1) but also the additional evidence associated with sets overlapping with U_1. When focal elements consist of class label singletons, belief and plausibility measures become equivalent.

To implement an evidential classifier, it is important to formulate a measure of evidence for subsets of class labels. Assume that focal elements consist of class label singletons. Then, measures of evidence for individual class propositions are represented as class membership values. However, the situation becomes more complex when mixed classes are involved, which are typical of remotely sensed images in highly varied terrain. Nonetheless, the capacity of evidential classifiers for handling mixed classes is well worth exploring. A quick follow-up in the context of multi-source data will be given in the next section, exemplifying the evidential approach for modelling mixed classes.

6.2.3 Combining Probabilistic Categorical Data

It has been shown how probabilistic categorical data may be derived from photo-interpretation and the computerised classification of remote sensor data. But the derivation of these data was implicitly cast in terms of separation. While geostatistical models can be inferred from photo-interpreted observational data for Kriging-based probabilistic mapping, probabilistic maps derived from remote sensor data with a chosen classifier are driven by spectral data where spatial dependence is usually not taken into account. How to combine geostatistically modelled and data-driven classifications in a theoretically sound manner presents an important topic for research, with the potential for results that are beneficial to information processing with multi-source data. In a GIS context, data from different sources may be correlated, although, in many cases, such correlations are not evaluated or made explicit. Spatial dependence is also a key characteristic for locations within a certain neighbourhood. The following section will investigate evidential, contextual, and geostatistical approaches to combining probabilistic data.

One approach to the combination of probabilistic data can be built upon evidential reasoning. Consider a location within the domain under study. Suppose that evidence for the location classification has been obtained from two sources that are assumed independent, such as spectral bands, and expressed as two basic probability assignments pm_1 and pm_2 whose associated classes are denoted by U_{k1} ($k1=1,\ldots,c1$) and U_{k2} ($k2=1,\ldots,c2$) respectively. The basic idea of Dempster's rule is that the two sources of evidence together yield the conclusion 'U_{k1} and U_{k2}' with probability $pm_1 pm_2$. The combination 'U_{k1} and U_{k2}' may be impossible, however, and the joint basic assignment of probability pm_{12} must be conditioned on such impossibilities. The two bodies of evidence can be combined, with a new set of classes denoted by U_k ($k=1,\ldots,c$), via summation over all possible pairs 'U_{k1} and U_{k2}':

$$pm_{12}(k) = \sum_{\text{possible pairs}} pm_1(U_{k1}) pm_2(U_{k2})/(1-K),\qquad(6.11)$$

which involves a factor K calculated from:

$$K = \sum_{\text{impossible pairs}} pm_1(U_{k1}) pm_2(U_{k2}),\qquad(6.12)$$

Mixed pixels that comprise heterogeneous ground conditions are known to pose a serious problem for remote sensing classification, which is designed to work with pure pixels of homogeneous categories, or pixels of inhomogeneous composition that can be resolved to their dominant categories. As a consequence, mixed pixels are rarely subjected to direct analysis in routine classification. However, it is possible to model the occurrence of both pure and mixed pixels by applying evidence theory, as Lee *at al.* (1987) discussed in comparing statistical and evidential approaches to the classification of remotely sensed data.

Suppose that a method of assignment is set up for two classes $k1$ and $k2$ and their mixture $k12$, which is to be applied to individual sources of evidence, and to reasoning based on their combination. Assume that the probabilistic membership vector for a location has been calculated from two sources of evidence, which yield (0.25,0.25,0.50) and (0.40,0.20,0.40) respectively. Clearly, the former probability vector indicates classes U_{k1} and U_{k2} are equally probable, and are most likely to be mixed, since the value assigned to U_k is the highest at 50%. This may suggest that the location would be better labelled as a mixture of the two classes rather than arbitrarily assigned to one class of the two. With the availability of another source of evidence, it is possible to interpret the latter probability vector as saying that class $k1$ is more probable than $k2$. But the probability of a mixed class is equal to the probability associated with the belief in class $k1$. However, using the rules for combining evidence, it is possible to calculate a combined probability vector, using the method shown in Table 6.1.

As shown in Table 6.1, the combined probability vector is (0.47,0.30,0.23), given the two sources of evidence concerning the probability distributions of alternative labelling of the location. This combined probability vector provides a relatively convincing suggestion that class U_{k1} be the most probable label for that location, although with a modest probability of 47%, as opposed to the cases using either source of evidence alone. This hypothetical example demonstrates that evidential classifiers lend themselves to building extensible and dynamic strategies for classification as extra evidence becomes available, and this ability of information updating and synthesis is very valuable. However, it is noteworthy that the evidential approach to combining probabilistic data relies on the availability of basic probability assignments, whose specification is far from trivial and remains a crucial issue.

Table 6.1 Evidential combination of probabilistic vectors:
Source 1 (0.25,0.25,0.50); Source 2 (0.40,0.20,0.40).

	pm_1	U_{k1}	U_{k2}	U_k
pm_2		0.25	0.25	0.50
U_{k1}	0.40	0.10	0.10	0.20
U_{k2}	0.20	0.05	0.05	0.10
U_k	0.40	0.10	0.10	0.20
$\sum pm_1\,pm_2$		0.40	0.25	0.20
$K = 0.15$				
pm_{12}		0.47	0.30	0.23

Geographical categories are often meaningfully discussed when they occupy spatial extents significantly larger than a single point. For this reason, spatial categories are often said to be context-dependent in the sense that they are distinguished from one another not in isolation but in neighbourhoods. This contextual dimension of categorical data is especially important in inferring high-level categorical data such as land use, which entails cultural as well as physical interpretations of land cover (Barnsley and Barr, 1996). Townsend (1986) used a logical smoothing operator based on majority filters for post-processing of per-pixel classified data, which greatly reduced the degree of the salt-and-pepper appearance typical of per-pixel classification.

Consider again the evidential approach to combined classification. If certain classified samples are available or recoverable from reference data and are well distributed, as is generally recommended, it is then possible to use them for creating geostatistical models of the set of categorical variables under study. From this, one can derive one set of probabilistic data, which are viewed as conveying contextual dependence and which can be combined with the probabilistic data derived from spectral data.

Switzer (1980) discussed contextual approaches to classifying remotely sensed images, in which reflectance data for a pixel are augmented with the averaged reflectance of its neighbours or other locally defined measures, taken as spatial contextual data, to drive discriminant analysis. The resulting classification was shown to be of improved accuracy as opposed to that using spectral data alone. In the image processing and pattern recognition communities, spatial contexts are often explored as sources of textural information. Carr (1996) researched various combined spectral and textural classifiers based on the minimum-distance-to-mean rule, by classifying Landsat TM data into surface cover types, such as shallow water in Yellowstone National Park, and demonstrated improved accuracy.

One may wish to analyse reflectance and, possibly, ancillary data by using techniques akin to cluster analysis and adding spatial constraints. The so-called contiguity-constrained spatial classification was devised to minimise variances within classes, by deriving spatially compact classes that are both similar in terms of measured features and spatially contiguous, and thus useful for planning and administration (Openshaw, 1977). Methods for classifying natural phenomena using spatial constraints were pursued by Oliver and Webster (1989), who investigated a geostatistically based and spatially constrained multivariate classification with soil survey data from two study areas in Britain. They calculated sample semivariograms from the original samples or their leading principal components, and fitted them to suitable models, which were then used to modify dissimilarity values by incorporating the form and extent of spatially dependent variation in the sampled soil variables, leading to improved spatial coherency in classified maps. This method for deriving spatially coherent classifications from the combined use of clustering analysis and geostatistics was developed further by Lark (1998). However, as is a persistent problem with any clustering-based classification, the resulting classes may not always make apparent sense, and human interpretation and intervention are necessary.

The above approaches to combining probabilistic data take into account the spatial dependence of class occurrence by exploiting either contextual information or the spatial dependence of feature (spectral) data, or both. Accommodating the dependence between data derived from different data sources and within spatial neighbourhoods is of great importance. Suppose two sources of data are available for probabilistic analysis of the occurrence of a class k at location x, denoted by $p1_k(x)$ and $p2_k(x)$ respectively. The probability for joint occurrence of the class at that location, given the two sources of data, will not simply be the product of $p1_k(x)$ and $p2_k(x)$, unless the two sources are assumed to be absolutely independent. The lineage of the data production process often suggests otherwise, because similar or identical methods, instruments, and sources may be used in both processes. If the probabilities from the two data sets are identical and class occurrences across the two sources are perfectly correlated, joint occurrence of the same class will have the same probability. The actual situation is likely to lie somewhere between these two extremes.

A similar argument for spatial dependence applies to a probabilistic pair $p_k(x_1)$ and $p_k(x_2)$ at spatially neighbouring locations x_1 and x_2. The joint probability for occurrence of class k at locations x_1 and x_2 will not simply be the product of $p_k(x_1)$ and $p_k(x_2)$, unless class allocation at the two locations is assumed to be entirely independent, a very unrealistic assumption for spatial data. It would be too restrictive to imagine perfect correlation among these two neighbouring locations such that their joint probability for class U_k remains the same as that at individual locations.

To develop strategies for utilising both spectral and spatial information and accommodating spatial dependence in probabilistic categorical mapping, co-Kriging may be helpful, using the indicator Kriged data and the spectral-data-driven probabilistic vectors. Suppose that two sets of probabilistic vector data are available in the form of $(p1_1(x_{s1}),...,p1_c(x_{s1}))$ $(s1=1,...,n1)$ and $(p2_1(x_{s2}),...,p2_c(x_{s2}))$ $(s2=1,...,n2)$, within a frame of c classes. The combined probabilistic estimate $p_k(x)$ for a location x is:

$$p_k(x)^* = \sum_{s1=1}^{n1} \lambda_{s1}(U_k;x)p1_k(x_{s1}) + \sum_{s2=1}^{n2} \lambda_{s2}(U_k;x)p2_k(x_{s2}), \text{ for } k=1,...,c \qquad (6.13)$$

where weighting parameters λ are solved by co-Kriging using a matrix of covariance and cross-covariance models inferred from both data sets (Myers, 1982; De Bruin, 2000a).

However, co-Kriging relies on stringent modelling of semivariograms and cross-semivariograms related to the indicator Kriged and the data-driven probabilistic fields, and its implementation may become extremely complicated, especially when the number of classes under consideration gets large. For this reason, indicator Kriging is usually implemented one category at a time, as discussed in Section 6.2.1, without making a fuller use of all the possible categories under study. Analysis is simplified by retaining only the spectral-data-based probabilities co-located with the individual grid nodes to be estimated, because this avoids the problem of modelling the spatial covariance of the probabilistically classified data, as has been seen in the case of continuous variables. A further approximation is achieved by building on a Markov-like hypothesis whereby co-located data are used to screen data of the same type located further away, so that the cross-covariance between the indicator data and the probabilistic classification has a linear relationship with the covariance of the indicator data (Almeida and Journel, 1994). The parameters for simplified spatial cross-covariance models are obtained through the availability of calibration data composed of indicators and their corresponding probabilistically classified spectral data.

6.2.4 Probabilistic Mapping of Land Cover

As stated in Section 5.4, mapping in suburban areas usually involves geographical data of a hierarchy of scales and accuracies. Such a scenario involves fine-scale map data on features such as buildings, contours, street networks, and land records on the one hand, and coarse-scale data such as land cover, soil, and noise levels on the other.

As a case study, part of the suburban area of the city of Edinburgh, near Blackford Hill, was again chosen as the test site, as in the previous chapter. This site contains a variety of topographic and thematic features, including a wooded valley, residential, commercial and academic buildings, roads, footpaths, a small lake, agricultural fields, and worked allotments. The residential areas consist of a complex mosaic of roads, pavements, roofs, walls, and hedges, which present a major challenge in deriving land-cover classification from remotely sensed data.

For the purpose of this study, the 1:24,000 scale colour aerial photographs, acquired in mid-June 1988 and already used in the test of terrain modelling, were used to derive probabilistic categorical data on the basis of photo-interpretation. For possible corroboration with photographic data, an extract of a Landsat TM image acquired over the same site in May 1988, virtually coincident with the 1:24,000 scale aerial photographs, was also used, as shown in Figure 6.3. This Landsat TM image extract was meant to complement probabilistic mapping with photo-interpretation, and its potential uses for driving various tests will be addressed in later sections as well.

Any spatial classification should reflect the resolution of the sensors collecting the attribute data, as outlined by Anderson *et al.* (1976) in their proposal for a four-level hierarchical land-cover classification designed with different sensors chosen to match the intended level of classification. This test focused on a five-class classification scheme, comprising grassland, built-up land, woodland, shrub, and water, and was geared to the combined use of aerial photographic and satellite image data.

Figure 6.3 An extract of a Landsat TM image of an Edinburgh suburb
(a summation of bands 3, 4, and 5 is shown).

Photo-interpretation was performed on the photographic stereo-model of the site, reconstituted using the photo control described in the previous chapter. Classified samples with a spatial extent equivalent to the Landsat TM pixel size (30 m by 30 m) were identified from locations where land cover could be considered comparatively pure or representative of the named classes. A set of 815 classified samples was collected from photo-interpretation, as shown in Table 6.2, and was used to drive indicator Kriging and as a source of reference data to validate subsequent analysis.

Table 6.2 Composition of classified samples.

Land cover types	Classified samples	
	numbers	percentages
Grassland	234	28.7
Built-up land	291	35.7
Woodland	203	25.0
Shrub	76	9.3
Water	11	1.3
Total	815	100.0

Each of these sample locations may be assumed to have a membership value of 1.0 in the named class, and consequently 0.0 for all other classes, resulting in binary encoded indicator data, as is usually the practice with indicator Kriging. These indicator data led to the semivariograms shown in Figure 6.4.

Shown in (a)–(e) of Figure 6.4 are semivariograms of indicator-encoded sample data for categories of grassland, built-up land, woodland, shrub, and water, respectively, with experimental and model semivariograms shown as diamonds and fitted curves. Clear between-class variabilities, in terms of ranges and sills, are evident in Figure 6.4.

Kriging was undertaken with the cell size of the output grids set to the 30 m spatial resolution of the Landsat TM sensor, utilising Kriging subroutines provided in GSLIB (Deutsch and Journel, 1998). This resulted in probability vectors at individual locations for the test site. Results are shown as surfaces in Figure 6.5, where (a), (b), (c), (d), and (e) illustrate probabilistic occurrences of grassland, built-up land, woodland, shrub, and water, respectively, and show clear variation over space.

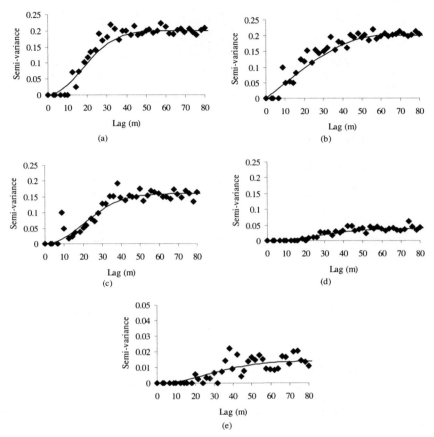

Figure 6.4 Experimental (diamonds) and model (fitted lines) semivariograms of indicator sample data: (a) grassland; (b) built-up land; (c) woodland; (d) shrub; and (e) water.

A suitable rule for evaluating Kriged probabilistic maps is to check for the consistency in the marginal distributions of individual classes between Kriged and sample data. Kriging generated probabilistic maps with average probabilities of 33.1%, 28.7%, 27.6%, 9.6%, and 1.0% for the five classes of grassland, built-up land, woodland, shrub, and water. It is clear by checking with the data listed in Table 6.2 that, on this test, built-up land was the least accurately mapped by Kriging, since the average probability of built-up land was estimated at 35.7% by the sample data and was under-estimated by about 20% at 28.7% in the Kriged map.

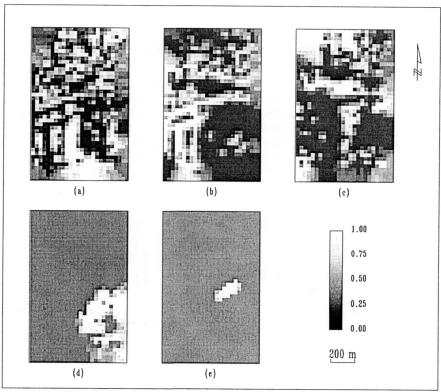

Figure 6.5 Probabilistic maps of land cover created from indicator Kriging:
(a) grassland; (b) built-up land; (c) woodland; (d) shrub; and (e) water.

The set of probabilistic maps of land cover derived from indicator Kriging was also used as the basis for probabilistic classification of the Landsat TM data, and as the reference to benchmark classification accuracy. Thus, a set of 336 classified samples was collected from the photogrammetric data. These samples were divided randomly into two sets, I (training) and II (testing), both comprising 168 pixels, as shown in Table 6.3.

Table 6.3 Compositions of training (I) and testing (II) pixels.

Land cover types	Numbers of pixels	
	I	II
Grassland	45	42
Built-up land	57	50
Woodland	41	48
Shrub	19	26
Water	6	2
Total	168	168

    Probabilistic categorical mapping with multispectral satellite data depends on
properly defined probability functions. In what follows each spectral band is
treated separately. For class U_k, the relative frequencies of digital number $z(x)$ at
pixel x can be read from the class-specific histogram and interpreted as
approximating the probability densities $\text{prob}(z(x)|k)$. Because of this, probabilistic
mapping of the photo-interpreted samples at locations x, in the form of $p_k(x)$, was
carried out using the Bayesian Equation 6.8, assuming equal *a priori* class
probabilities. Probabilistic membership values from different evidence sources,
that is, spectral bands, were combined using the Dempster-Shafer method shown
in Equations 6.11 and 6.12. The resulting maps are shown as surfaces in Figure
6.6, where probabilistic occurrences of grassland (a), built-up land (b), woodland
(c), shrub (d), and water (e) show interesting spatial variation.

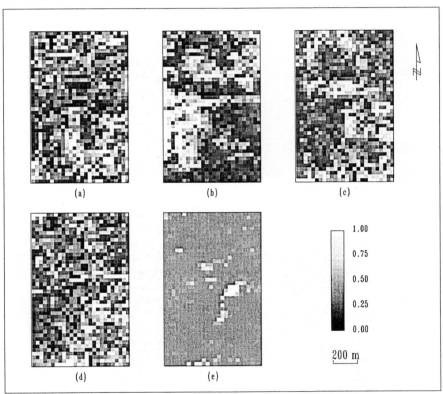

Figure 6.6 Probabilistic maps of land cover created from the Landsat TM data:
(a) grassland; (b) built-up land; (c) woodland; (d) shrub; and (e) water.

    To obtain conventional discrete categorical maps from probabilistically
classified data, an analogy with the classification of raster-based remotely sensed
images using the maximum-likelihood classifier is helpful. It is common practice
for maximum-likelihood classifiers to assigns pixels (locations) to the classes to
which they have the maximum likelihood of belonging, measured by the posterior

class probabilities defined in Equation 6.8. A straightforward extension of this algorithm is that categorical data in the form of readily available membership values are subjected to a maximisation process, by which location $x_s \in D$ is labelled with the class that has the maximum probability of occurrence.

A distinctive feature of probabilistic classification is that the probabilities of class occurrences may be retained rather than discarded as in discrete classification. Indeed, the magnitude of the maximum probability value can be used to generate a companion surface showing the spatial distribution of the certainty in class allocations (Zhang and Kirby, 1999). Categorical maps processed like this are represented by both discrete class allocation and information on the uncertainty pertaining to such classification. Classified maps and their corresponding surfaces of maximum class probabilities are shown in (a) and (b), (c) and (d) of Figure 6.7, for probabilistic maps derived from Kriging and evidential reasoning, that is, those depicted in Figures 6.5 and 6.6, respectively.

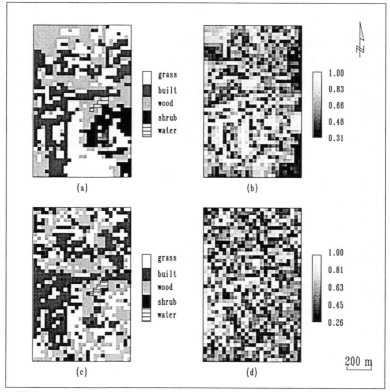

Figure 6.7 Maps of discrete classification and maximum class probabilities derived from: (a) and (b) Kriging of photo-interpreted data; and (c) and (d) probabilistic classified Landsat TM data.

Clearly, classifications derived from probabilistically classified Landsat TM data exhibit more variation than those based on Kriging. Their accuracy may be evaluated in terms of the PCC and the Kappa coefficient of agreement, and should be undertaken in a statistically sound manner (Congalton, 1991; Stehman and Czaplewski, 1998). The accuracy of the hardened probabilistically classified Landsat TM data was assessed with reference to the test data tabulated in Table 6.3. This resulted in a PCC of 51.2%, which is not at all satisfactory for practical use and suggests the need for further work.

With probabilistic categorical data, majority filtering may be conducted by averaging class probabilities within a local window. Such locally averaged class probabilities can then be reduced to discrete classification prior to accuracy assessment. Thus, post-processing of probabilistically classified Landsat TM data by majority filtering with a 3 by 3 window resulted in a PCC of 54.8%, hardly a significant improvement over the initial PCC of 51.2%.

Spatial classification is different from classifying non-spatial elements in the sense that the locations to be classified are rarely independent. Spatial dependence should therefore be analysed to gain a fuller description of the probable occurrences of underlying categories. To explore spatial dependence in categorical occurrences embedded in discrete training samples and probabilistically classified spectral data, covariances between training data and the probabilistically classified Landsat TM data shown in Figure 6.6 were analysed using GSLIB subroutines.

To facilitate data analysis, training samples were organised in a manner compatible with the geostatistical software system GSLIB. For this, sample data were organised so that each record was comprised of grid coordinates, and binary codes (0 or 1) for the five types of land cover under consideration. Pixels co-located with training samples were extracted from probabilistically classified Landsat TM data, with probabilistic vectors arranged with indicator-transformed data to support the calculation of experimental semivariograms and covariances. The results are shown in Figure 6.8, where experimental semivariograms are represented by diamonds and model semivariograms by curves.

Based on this, spatial contextual data were extracted from the training data via indicator Kriging on the basis of the semivariograms shown in (a), (d), (g), (j), and (m) of Figure 6.8. Contextual data, which were expected to enhance the accuracy of the probabilistically classified Landsat TM data, were initially treated as an extra independent source of evidence, with which the probabilistically classified Landsat TM data shown in Figure 6.6 were combined. This processing of contextual information gave rise to a new probabilistic classification of the Landsat TM data with a PCC of 74.4%, significantly higher than that obtained either with spectral data alone or with the results of post-processing using majority filtering. This reinforces the superiority of geostatistical methods for post-processing of classified data over simple filtering, where the spatial dependence embedded in the training data is utilised only in the assignment of discrete class labels.

With reference to the definition of probabilistic class memberships in Equation 6.8, which is based on histogrammed relative spectral frequencies, the foregoing analysis has implicitly assumed equal prior probabilities of class occurrences. But it is possible to use Kriged results from training data as location-specific and thus spatially varying prior probabilities of class occurrences. These prior data were then incorporated in the calculation of probabilistic class memberships, generating the probabilistic maps shown as surfaces in Figure 6.9.

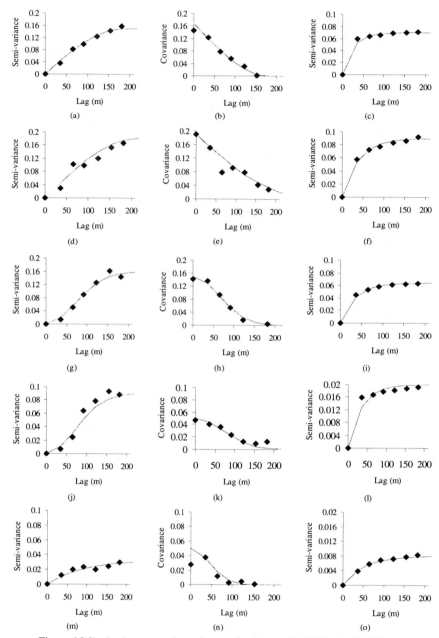

Figure 6.8 Semivariograms and covariances of training and initially classified data:
(a), (b), and (c) grassland; (d), (e), and (f) built-up land; (g), (h), and (i) woodland;
(j), (k), and (l) shrub; and (m), (n), and (o) water.

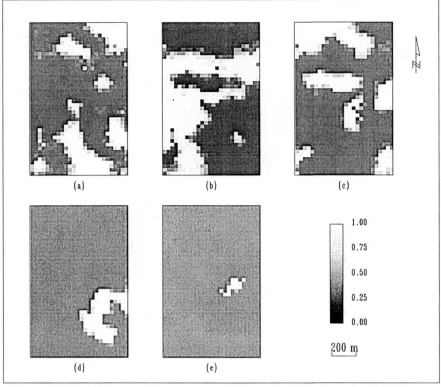

Figure 6.9 Probabilistic maps of land cover created from contextually enhanced probabilistically classified Landsat TM data: (a) grassland; (b) built-up land; (c) woodland; (d) shrub; and (e) water.

The probability surfaces of class occurrences, shown in Figure 6.9, look much smoother than those in Figure 6.6. Classification accuracy for this set of probabilistic maps tuned with prior data was 79.2%, even higher than that reported in the case of contextual classification. A plausible explanation is perhaps that the latter treated Kriged probabilistic data as an independent body of evidence for evidential reasoning, without explicit accommodation of the spatial dependence in probabilistically classified Landsat TM data, apart from that in the training data.

The direct influence exercised by training data upon the resulting classified maps may explain the higher agreement obtained by incorporating the spatial dependence implied in the training data than from classification based on spectral data alone. Ordinary Kriging using spectrally based probabilistic maps to estimate drift may also be helpful, as this may allow for conditioning to both spatial dependence and spectral information. An accuracy assessment of post-processed Kriging classification indicates a PCC of 67.8%, which is lower than the values obtained with contextual enhancement and the incorporation of prior data in probabilistic class definition, but higher than classifications based on spectral data or simple filtering alone.

With the semivariograms modelled in Figure 6.8, co-Kriging appears to provide interesting potential. The semivariograms and cross-semivariograms obtained from training data and probabilistically classified Landsat TM data were used to define correlation coefficients, which, in turn, enabled implementation of co-Kriging with co-located data, a specific procedure in GSLIB based on completely gridded secondary data to augment sparsely sampled primary data. The resulting probabilistic maps are shown in Figure 6.10, where probable occurrences of grassland, built-up land, woodland, shrub, and water are shown as surfaces in (a)–(e), respectively.

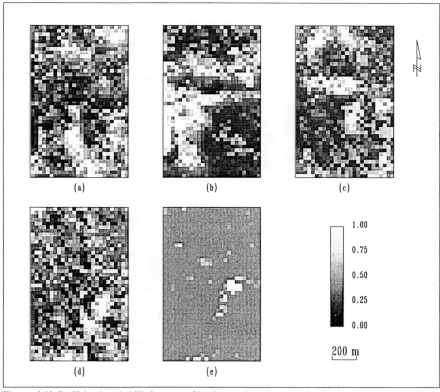

Figure 6.10 Co-Kriged probabilistic maps of land cover by using a probabilistically classified Landsat TM image as co-located data: (a) grassland; (b) built-up land; (c) woodland; (d) shrub; and (e) water.

Accuracy assessment of the classification using co-Kriging indicates a PCC of 61.3%, which is lower than that based on ordinary Kriging using spectrally classified data to estimate drift. This is because co-Kriging utilises both the spatial dependence implied by the training data and spectral-data-based classification, with the former taking precedence over the latter. As the training and testing data are drawn from domains that are not necessarily well mapped by spectral data, a stronger degree of conditioning by spectral data leads to lower agreement with the test data.

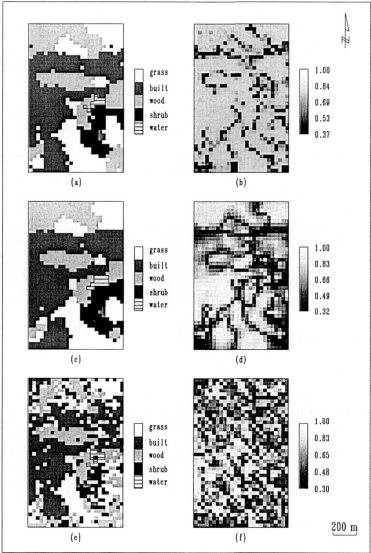

Figure 6.11 Maps of discrete classification and maximum class probabilities based on:
(a) and (b) post-processing of probabilistically classified Landsat TM data using geostatistical priors;
(c) and (d) Kriging using spectral classification to estimate drift; and
(e) and (f) co-Kriging using spectral classification as co-located data.

As before, probabilistic maps derived from contextual enhancement by a combination of spectral and spatial evidence, ordinary Kriging (Figure 6.9), and co-Kriging (Figure 6.10) may be reduced to a combination of discrete classification and probability pertaining to that class allocation. Results are reported in Figure 6.11, where the pairs (a)–(b), (c)–(d), and (e)–(f) represent classified maps and probability surfaces for contextual enhanced, ordinary Kriging, and co-Kriging post-processed classification of Landsat TM data.

As shown in Figure 6.11, post-processed classification by Kriging shows more evidence of spectral information in probabilistic class distributions than in those modified with geostatistical priors extracted from training data. In particular, the co-Kriging strategy for post-processing classified data utilises spectral information in spatial classification but without sacrificing the information on spatial dependence implicit in the training data. Because spatial classification involves a combination of subjective judgment and interpretation, as in the collection of training data, and objective measurements acquired by remote sensors, the geostatistics of spatial categories will provide a feasible solution to the optimal weighting of contextual and spectral data. Towards this end, co-Kriging has shown promise for expanding our understanding of spatial categories and their classification, through the extended use of spatial knowledge and spectral measurement. For this reason, post-processed spectral classifications by co-Kriging should be considered as a geostatistically balanced compromise between reliance on geostatistical models inferred from training data, and reliance on the spatial variation of the spectral signatures of candidate classes. Accordingly, the notion of ground truth for categories should be modified to reflect the geostatistical combination of human interpretation and remote sensor measurement. A similar idea was discussed by De Groeve *et al.* (1999).

6.3 PROBABILISTIC MODELLING OF CATEGORIES

6.3.1 Probable Categorical Occurrence and Stochastic Simulation

Probabilistic categorical mapping based on ground survey and remote sensing represents an improvement over conventional methods, because the former retains more information on the stochastic occurrence of individual classes than the latter, where a deterministic image is rendered of class occurrence over space. This is further reinforced by combining geostatistical models with data-driven probabilities for better estimates of the uncertainties intrinsic to spatial classification. Further analysis of categorical data can lead to an assessment of the reliability of using uncertain categorical data for spatial query and decision-making.

Consider the combination of spatial data involving categorical variables. Newcomer and Szajgin (1984) found that map overlaying using AND operations tends to degrade the accuracy of the derived map products as the number of overlaid layers increases. Veregin (1989) extended the error propagation model to both logical AND and OR operations. In contrast to AND operations, OR operations tend to inflate the accuracy in the derived map products. While these experiments are reasonably sound and constitute valuable research contributions to GIS uncertainty, it is possible to advance further based on the logic of categorical data. As in uncertainty modelling with continuous variables, Kriging is rarely enough to provide case-specific estimates of spatial uncertainties in Kriged maps, which are, in a sense, averaged estimates of spatial distributions. Thus, even if probabilistic categorical data are available, it may not be possible to use them alone to elucidate unbiased estimates of spatial uncertainties in derived data products, as there is a persistent need to deal with spatial dependence.

Reporting and statistical analysis of categorical area estimates from spatial databases are common functions of GISs. Remote sensing provides computerised techniques for deriving categorical maps via image classification (Bauer *et al.*, 1978). Discrete categorical maps serve such purposes readily by assuming deterministic data representation, giving statistics that are also crisp. With probabilistic categorical data, it is possible to get most probable statistics on spatial categories, and hence to improve the accuracy of areal estimates.

Suppose that a set of probabilistic categorical maps, $\{p_1(x),...,p_c(x) \mid x \in D\}$ are constructed from sample data in order to describe spatially varying levels of confidence in class occurrences within a field domain D. Let a be the area of a raster cell. The mean total area belonging to class U_k ($k=1,...,c$) is the sum of $p_k(x)$ multiplied by a^2, given that $p_k(x)$ is the estimated mean probability of occurrence of class U_k. The question of uncertainty regarding the total area estimate is, however, not answerable simply by calculating binomial distribution variances by summing the product $p_k(x)(1-p_k(x))$ multiplied by a^2 for class U_k across the field domain, unless zero spatial dependence is assumed. Many spatial categories are shown to exhibit positive spatial correlation over varied neighbourhoods, which implies that variance in areal estimation is likely to be underestimated using the binomial distribution model (De Bruin, 2000a).

Another issue is related to the complexity of spatial categorical reasoning with joint logic. In fact, many real-world applications are not defined sufficiently, and simple arithmetic or logic operations on probabilistic data rarely suffice to drive reasoning about uncertainty. For instance, it is not clear what would be the reliability of using both continuous and categorical variables (e.g., terrain elevation and land cover) with given estimates of uncertainties to support a query such as the probabilistic occurrence of alpine against broadleaf forest within certain elevation ranges. Clearly, uncertainty-informed decision-making is not sufficiently supported by the estimation of uncertainties in source data, but requires instead the use of realisations of equal-probable component map layers, which can then be used to drive specific map overlay rules.

Lastly, in certain situations, it is simply not easy to derive probabilistic data in the first place, especially when those data contain inherent subjective elements. This section aims to take a step beyond probabilistic classification, and to satisfy the practical need to assess the consequences of using uncertain categorical variables in GIS problem-solving, within the context of heterogeneous data. A good way forward is to view categorical maps as equally probable realisations of certain stochastic processes (Goodchild, 1994).

Uncertainty models can be conceptualised as stochastic processes capable of generating distorted alternatives of an underlying reality. A model suitable for generating equal-probable categorical data should have two essential properties. Between realisations, the proportion of times pixel x is assigned to class U_k should approach $p_k(x)$, as the number of realisations becomes large. The other property requires that, within realisations, the outcomes in neighbouring pixels be correlated to a degree specified by spatial dependence parameters. When the spatial dependence parameter is set to zero, class occurrences in neighbourhoods are independent of one another, an idealised situation pursued by Fisher (1991) using Monte Carlo simulation. As the degree of spatial dependence becomes increasingly positive, outcomes are correlated over longer and longer distances, with larger and larger anomalous inclusions being ignored within polygons or falling below the minimum mapping unit area (Goodchild *et al.* 1992).

A method for categorical simulation is through the use of autoregressive models, which were discussed in the previous chapter as a basis for simulating continuous variables. For modelling uncertain categorical data, per-class probabilistic distributions are assumed to be available. A potential user must provide specific parameters of spatial dependence in order to run the procedure to generate realisations conforming to his or her knowledge about the degree of spatial correlation. This process for parametric specification is somehow less appealing to users who may be worried about the subjectivity involved. A more user-oriented and also commonly used algorithm for simulating categorical variables is sequential indicator simulation (Goovaerts, 1996), which can be carried out in both unconditional and conditional forms.

Consider a domain D where discrete categories $\{U(x), x \in D\}$ are to be simulated. Suppose n classified samples exist in the form $\{u(x_s), s=1,...,n\}$, where $u(x_s)$ are realisations of c mutually exclusive categories U_k of a prescribed ordering $(k=1,...,c)$. It is then straightforward to transform each categorical sample into a vector of c indicator data using the definition given in Equation 6.1. By defining a random path traversing each node to be simulated, the sequential simulation proceeds as follows:

1) At each node x, determine the local probability of occurrence of each category U_k, $p_k(x \mid \text{data})$, using indicator Kriging. The conditioning data consist of neighbouring original indicator data and previously simulated indicator values, the former being withheld for unconditional simulation. Simulation conditional to a combination of indicator data and ancillary data may be implemented with approximation by retaining only ancillary data co-located at the grid nodes being simulated, and simplifications built upon the Markovian hypothesis described by Almeida and Journel (1994).

2) Normalise so that simulated values are between 0 and 1, and sum to one.

3) Build a cumulative distribution function (cdf) using:

4) $$F\big(U_k; x \mid \text{data}\big) = \sum_{j=1}^{k} p_j\big(x \mid \text{data}\big), k = 1,...,c \tag{6.14}$$

5) Draw a random number p uniformly distributed between 0 and 1. The category simulated at location x, $u^1(x)$, is the category that corresponds to the probability interval including p, that is:

6) $u^1(x) = U_k$ if $p \in \big\{\big(F(U_{k-1}; x) \mid \text{data}\big), \big(F(U_k; x) \mid \text{data}\big)\big\}$ $\hspace{1cm}$ (6.15)

7) Add that simulated value to the conditioning data set, and continue the previous four steps for the next node along the random path.

Upon the complete traversal of all the nodes to be simulated, a realisation $\{u^1(x), x \in D\}$ is obtained. The same procedure is repeated with a different random path to generate further realisations $\{u^\tau(x), x \in D\}$, $\tau \neq 1$.

Continuous variables such as terrain elevation and slope gradient have been shown to depend on the quantification of spatial dependence for objective uncertainty modelling. Basing analysis of uncertainty in categorical occurrences on the amount of spatial dependence is equally important, as the next section explains.

6.3.2 Modelling Uncertain Land Cover

Kriged probabilistic maps such as those shown in Figures 6.9 and 6.10 provide information on spatially varying uncertainty in class allocation. But they are not sufficient for deriving estimates of uncertainty in spatial queries involving the joint occurrences of spatial categories, such as estimates of area or map overlay. Stochastic simulation is required for modelling uncertainties in categorical data and spatial applications involving spatial categorisation, if the purpose is more than mere description of uncertain occurrences.

With the classified samples collected as in Table 6.3 and the semivariograms derived in Figure 6.8, it is possible to pursue the simulation of land cover data. Both unconditional and conditional modes of simulation were performed, each with 100 realisations. It is understood that unconditional simulation applies specified semivariogram models, while the conditional alternative honours, in addition, classified sample observations and co-located spectrally classified data in generating spatially correlated categories. Results are shown in Figure 6.12, where three example realisations are depicted for (a)–(c) unconditional and (d)–(f) conditional simulated maps, respectively.

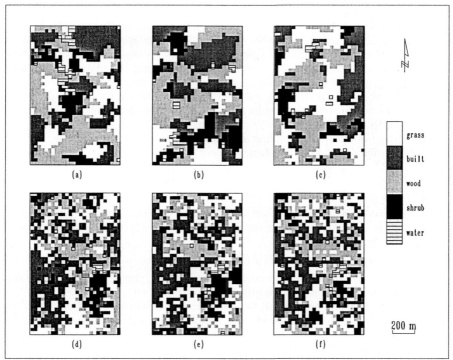

Figure 6.12 Stochastically simulated maps of land cover:
(a)–(c) unconditional realisations; and (d)–(f) conditional realisations.

As one would expect, both unconditional and conditional modes of categorical simulation result in substantial differences between individual realisations regarding the spatial distributions of categories, except at known locations in the case of conditional simulation. The simulated categorical maps shown in Figure 6.12 reveal, also, the distinctive properties of unconditional and conditional approaches to simulating spatial categories. For example, unconditionally simulated maps appear less varied than conditionally simulated ones, as the latter are tuned with location-specific categories and spectrally classified data, which are bound to exhibit much spatial variability.

Simulated categorical maps are useful as a basis for deriving both probabilistic maps and conventional categorical maps of discrete classes, together with visualisations of spatially varying uncertainties in class allocation. The results are summarised for unconditional and conditional simulated realisations, and are shown as probabilistic maps in Figures 6.13 and 6.14, respectively.

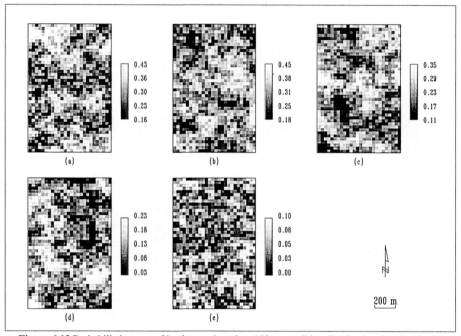

Figure 6.13 Probabilistic maps of land cover based on 100 unconditionally simulated realisations: (a) grassland; (b) built-up land; (c) woodland; (d) shrub; and (e) water.

With probabilistic mapping, as alternative realisations are summarised, local variations between realisations, like those shown in Figure 6.12, will be smoothed to some extent. This characteristic is clearly visible in Figures 6.13 and 6.14. Probabilistic maps derived from simulated categorical maps can be further processed to produce discrete representations of land cover and surfaces of probability, similar to the case with Kriged probabilistic maps. Results are reported in Figure 6.15, where classified land cover maps and the corresponding maps detailing spatially varying certainties in class allocation are shown as pairs (a)–(b) and (c)–(d) for unconditionally and conditionally simulated data, respectively.

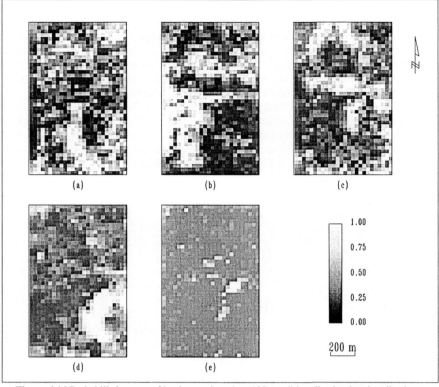

Figure 6.14 Probabilistic maps of land cover based on 100 conditionally simulated realisations: (a) grassland; (b) built-up land; (c) woodland; (d) shrub; and (e) water.

The maps shown in Figure 6.15 display a notable difference between summarised categorisations based on unconditional and conditional simulation. With unconditionally simulated categorical maps, categorisation based on the most frequent occurrence at individual locations tends to be dominated by the categories of grassland and built-up land, with woodland severely underestimated while shrubland and water bodies are totally missing. On the contrary, categorisation based on conditionally simulated categorical maps retains the observed and highly probable categories, based on deterministic indicators and spectrally classified data.

Simulated categorical maps are believed to behave in a way close to the underlying reality. It has been seen that Kriging produces smoothed maps of classification. Similarly, a summation of simulated maps will show greatly reduced local variation in categorical distributions, as shown in Figures 6.13 and 6.14. Simulation allows for reproduction of the categorical heterogeneity and variability originally prevailing in the underlying problem domain, allowing complex spatial query and decision-making to be assessed in an uncertainty-informed way. Generation of probable categorical maps is a major step forward. But, to make fuller use of stochastic simulation, complex spatial queries such as joint class occurrences rather than simple problems, such as categorisation in individual locations, must be supported. Areal estimation is one such problem involving the joint occurrences of classes over space and in neighbourhoods.

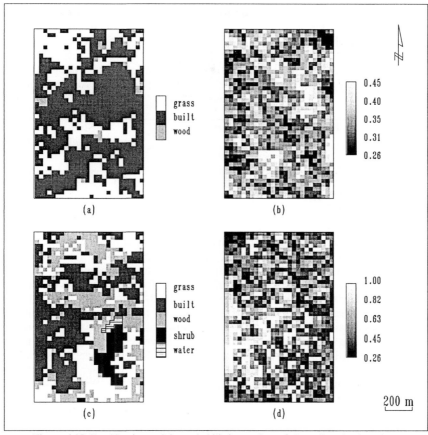

Figure 6.15 Classification and the probabilistic certainty of class allocation based on: (a)–(b) unconditional simulation; and (c)–(d) conditional simulation.

Area estimates of different categories are standard outputs from GIS. As was indicated in Section 6.3.1, mean estimates of categorical areas should be improved by summing probabilistic data across individual locations, as categorical proportions in individual locations of finite size can be mapped by indicator Kriging, leading to improvement over the counting of discrete classes. Binomial estimates of standard deviation for class probabilities are, however, biased for categorical data containing significant spatial correlation. To provide empirical evidence regarding the relevance and significance of incorporating spatial dependence in categorical simulation for statistically sound estimation of uncertainties about area statistics, a program was written importing simulated map data and subjecting them to statistical analysis.

Quantitative results are reported in Table 6.4, where per-class estimates, in units of Landsat TM pixels (i.e., 30 m by 30 m), of mean areal extents and their standard deviations are tabulated for those derived from stochastic simulation, as well as from the Kriging documented earlier in Section 6.2.4. The reference data that are listed along with Kriged and simulated data are based on the categorical map shown in Figure 6.7(a), which was derived from indicator Kriging of 815 discrete categorical samples.

Table 6.4 Comparison of areal estimates and their errors based on Kriged and simulated data (except for 'Reference', the left and right values refer to means and standard deviations respectively, in units of Landsat TM pixels).

Land cover types	Reference	Kriging crisp		Kriging probable		Simulation unconditional		Simulation conditional	
Grassland	355	292	15	300	12	313	68	301	19
Built-up land	317	397	16	343	11	365	79	370	17
Woodland	305	274	14	252	11	242	84	273	16
Shrub	107	118	10	185	11	131	53	131	18
Water	8	11	3	12	2	40	19	15	3

Examination of Table 6.4 shows that no one method holds a clear advantage over the others in terms of the closeness of mean categorical area estimates to reference values. However, stochastic simulation seems to give more consistent results across land cover types than Kriging, except for the cases of woodland and water bodies. For the Kriged maps, except for built-up land, crisp classification gives rise to areal estimates that are closer to reference values than does probabilistic classification, which retains information on spatially varied class memberships. The most striking discrepancy is the over-estimation of built-up land by Kriging-based crisp class allocation. Between different modes of simulation, conditional simulation gives rise to areal estimates that are closer to the reference values than their unconditional counterparts, for all land cover types except for grassland and built-up land.

As is shown by Table 6.4, the standard deviations in areal estimates by simulation are much greater than those evaluated by Kriged maps, and the results of unconditional simulation indicate variances that are about five times those obtained by Kriging. Standard deviations in areal estimates by crisp classification are higher than those from probabilistic classification, while conditional simulation leads to standard deviations in areal estimates slightly greater than those from Kriging. As can be anticipated, conditional methods for simulating spatial categories, in which observed class occurrences are honoured, lead to greatly reduced variances in areal estimates as opposed to unconditional ones. Thus, it is suggested that stochastic simulation can be used to achieve greater accuracy in categorisation, in addition to its flexibility in accommodating the spatial dependence that is intrinsic to occurrences of many spatial categories.

It should be recognised that, whether Kriging or stochastic simulation is applied for categorical estimation, the resultant statistics will be as accurate as how well the categorical samples are collected in the first place. In other words, accuracy in categorical mapping will be dependent of not only the specific estimation methods chosen but also the sampling design, as remarked by Longley *et al.* (2001). Clearly, the data listed in Table 6.4 should be interpreted in the context of the sampling underlying Tables 6.2 and 6.3.

6.4 CATEGORICAL FUZZINESS

6.4.1 Fuzzy Sets and Classification

As has been shown previously in this chapter, a fundamental assumption in the realm of conventional categorical mapping is that the spatial categories prevailing at individual locations are unambiguously defined at a chosen scale. In other words, there exists conceptual truth regarding categorical occurrences at individual locations, and true classes are recoverable given sufficient data and knowledge. Through remote sensing techniques, discrete categories are identified on the basis of spectral data (e.g., reflectance), ancillary data (e.*g*., topography) if available, and ground survey. Spatial categories are treated as crisp sets, with each individual location exactly identified with a single category.

From an empirical perspective categorical mapping is far from certain, and various inaccuracies occur. Above all, categorisation of inherently complex spatial phenomena into discrete classes is bound to involve subjectivity in such areas as class description and sample selection. Observational data are rarely adequate for perfect class labelling. The combination of subjectivity and inherent data limitations leads to varying uncertainty in the categorical data obtained through this process. Misclassifications emerge, when disagreement is revealed between what is meant to be true and what is observed and inferred, typically by checking experimental classified data against a reference data set. Thus, a location is either correctly classified or totally misclassified, amounting to a realisation of crisp binary sets.

To describe such inaccuracy with categorical mapping, one may adopt a probabilistic formalism for a classified map by estimating an averaged percentage of correctly classified locations. In addition to this global estimate, it is possible to conceive of uncertainties in class allocation as being spatially varying, as seen in the previous section. At this spatially explicit level, a probabilistic description of class occurrences at a location can be created from geostatistical tools if classified samples are available. Although class occurrences are now interpreted probabilistically, crispness is still embedded in the process of collecting classified samples, and the discrete representation of probabilistically encoded maps, and in equal-probable realisations of stochastic simulation.

While the probabilistic formalism is useful for accommodating categorical uncertainties due primarily to randomness, there are many other factors contributing to uncertainty, which are not always well described and modelled by probability. As widely admitted, categorical data often suffer from inherent conceptual inexactness and vagueness, which, in a geographical context, are often worsened by complexity. GIS applications involve many spatial categories that exist as graded continua, are inherently vague, and should not be considered as being governed by or exhibiting crispness. Consequently, the applicability of probability theory and its underlying crisp set theory is rendered unsuitable by vague categories.

Various types of spatial categories are vaguely defined and have poor areal definition at all scales. For example, while grassland and woodland are conceptually distinct, the categorisation of woodland into areas of close and open canopy, or grass into low- and high-growing subtypes is inherently vague. So is the concept of vegetation abundance. Similar examples for fuzzy classes exist in dryland salinity monitoring and assessment, where vagueness is evident in a categorisation between low, moderate, and severe salinity. Finally, atmospheric pressure, cloud thickness, rainfall intensity, and urbanisation may each be described inexactly as low, moderate, or high. These examples highlight common causes of vagueness, which is related to conceptual gradation between classes, similar to the situation with categories measured on the ordinal scale. In this section, fuzziness is used as a term subsuming all types of inexactness and vagueness, given the literature accumulated over past decades on fuzzy classification. Fisher (2000) provided a recent account for the legacy of vague geographical information.

Thus, fuzzy sets seem to be a natural choice for expressing the vagueness that is intrinsic to many spatially distributed categories. It is useful to exploit and explore the potential advantages of fuzzy set theory, and associated methods developed elsewhere in the academic world (Zadeh, 1965; Bezdek *et al.*, 1984). Under fuzzy set theory, the transition between membership and non-membership is gradual, and any location (pixel) x belongs to fuzzily defined classes U_k ($k=1,...,c$) with a membership vector $\mu_k(x)$ ($k=1,...,c$) conveniently valued within the unit interval $[0,1]$. Further, fuzzy membership values for a location typically sum to 1.0 across all classes. It is possible to conceive of spatial distributions of fuzzy class memberships as continuous-valued fields, which depict spatially varying strength of fuzzy belonging, using a format similar to that shown in Figure 6.1(b).

The conceptual superiority of fuzzy spatial classification has been demonstrated in many studies. Fuzzy classification avoids problems associated with conventional crisp or hard classification and has become a popular alternative method for categorical thematic mapping by remote sensing (Wang, F., 1990; Foody, 1996). For example, statistical and neural approaches to fuzzy classification have been widely used for mapping spatial continua of land cover and soil types with varied success (Foody, 1995; Zhu, 2000). In these studies, graded memberships were estimated usually by employing representational models for similarity, which were able to estimate the fuzzy membership of a mapping unit in a prescribed set of classes, or the deviation of a mapped unit from the ideal prototype of its nominal class.

Fuzzy set theory has also been explored for possible application in the accuracy assessment of fuzzy classified data. A fuzzy approach to assessing classification accuracy has also been researched using fuzzy sets constructed on a linguistic scale, the classical interpretation for fuzzy sets (Zadeh, 1984), which allows the kinds of comments commonly made during map evaluation to be used to quantify map accuracy (Gopal and Woodcock, 1994). The use of fuzzy sets in accuracy assessment expands the amount of information provided regarding the characteristics and sources of categorical uncertainties beyond those possibly interpreted from conventional error matrices (Woodcock *et al.*, 1996; Jager and Benz, 2000).

A well-known approach to fuzzy classification is the fuzzy c-means clustering algorithm (Bezdek *et al.*, 1984). Let $Z(x)=\{z(x_1),...,z(x_n)\}$ be a sample of

n elements (pixels), each attributed with vector observations (multispectral measurements) in an nb-dimensional Euclidean space (e.g., with nb spectral bands). A fuzzy clustering, with reference to n pixels comprising the field domain and c clusters, is represented by a fuzzy set $\{\mu_k(x_s) \in [0.0, 1.0]\}$ ($s=1,...,n$; $k=1,...,c$), where $\mu_k(x_s)$ is the fuzzy membership value of an observation $z(x_s)$ in the kth cluster. A fuzzy membership value ranges between 0.0 and 1.0 and is positively related to the degree of similarity or strength of membership of an element in a specified cluster.

A variety of algorithms has been developed for deriving an optimal fuzzy c-means clustering (Ruspini, 1969). One widely used method works by minimising a generalised least-squared error function J_m:

$$J_m = \sum_{s=1}^{n} \sum_{k=1}^{c} \left(\mu_k(x_s)\right)^m \left(Mdis_{sk}\right)^2 \tag{6.16}$$

where m is the weighting exponent which controls the degree of fuzziness (increasing m tends to increase fuzziness; usually, the value of m is set between 1.5 and 3.0), and $Mdis_{sk}^2$ is the squared Mahalanobis distance between each observation $z(x_s)$ and a fuzzy cluster centre (m_k), calculated using Equation 6.7 (Bezdek *et al.*, 1984).

The minimisation of the error function J_m begins by setting μ_k randomly. An optimal fuzzy partition is then sought in an iterative way. Thus, fuzzy c-means clustering is often applied in an unsupervised mode. It is important to note that the mean vectors in cluster analysis that define class prototypes may not have an exact match within the population, suggesting vagueness in classification. Since the error function is globally minimised, the resulting clusters may be visually unappealing. However, entropy can be used to evaluate the degree of fuzziness of fuzzily classified data, thus aiding in the validation of fuzzy clustering. Entropy $H(c)$ for a set of fuzzy classified pixels is measured as:

$$H(c) = -\sum_{s=1}^{n} \sum_{k=1}^{c} \mu_k(x_s) \log_2 \mu_k(x_s) \tag{6.17}$$

where $\mu_k(x_s)$ is the fuzzy membership value of grid cell x_s in class k, where the index s ranges from 1 to n (the total number of classified pixels), and where k ranges from 1 to c (the total number of classes).

Measures of entropy express the way in which the class memberships are partitioned between the classes, based on the assumption that in an accurate classification each location will have a high membership in only one class. As entropy is a measure of disorder, large values indicate a low accuracy of classification, while small values indicate high accuracy. Thus, a plot of H against c may be examined for local minima to obtain an optimal value of c.

The algorithm outlined above for fuzzy clustering may be modified for the derivation of fuzzy classification in a supervised mode. For this, the class centroids (v_k) are determined from the training data. This reduces the algorithm for fuzzy c-means clustering to a one-step calculation, resulting in a fuzzy membership value for each pixel in each of the defined classes. The fuzziness of the classification can be modified by varying the magnitude of the parameter m. With a given value of m, fuzzy memberships may be adjusted by using per-class covariance matrices rather than global ones. Such a strategy based on per-class means and covariances allows fuzzy c-means clustering to be adapted easily to a supervised implementation using the same fuzzy training data.

Suppose that the spectral data of the training pixel set are $\{z_1,...,z_n\}$, and the fuzzy training data are represented as a fuzzy set $\{\mu 0_k(x_s)\}$ with reference to n training pixels and c clusters. The element $\mu 0_k(x_s)$ represents the fuzzy membership value of a training pixel z_s $(1\leq s\leq n)$ in the kth cluster. Using the fuzzy training data, it is possible to define a fuzzy mean (M^*_k) and a fuzzy covariance matrix (COV^*_k) for fuzzy classification (Wang, F., 1990; Zhang and Foody, 2001) using the notions of fuzzy expectation and covariance (Zadeh, 1968; Viertl, 1997). The formulae for calculating fuzzy means and covariances are:

$$M^*_k = \sum_{s=1}^{n} \mu 0_k (x_s) z_s \left/ \sum_{s=1}^{n} \mu 0_k (x_s) \right. \tag{6.18}$$

and:

$$COV^*_k = \sum_{s=1}^{n} \mu 0_k (x_s)(z_s - M_k)(z_s - M_k)^{\mathrm{T}} \left/ \sum_{s=1}^{n} \mu 0_k (x_s) \right. \tag{6.19}$$

The success of fuzzy classifications depends on how well membership functions are defined, a rather subjective process. It has been shown that fuzzy membership functions can be derived from continuous-valued measurements such as spectral reflectance via fuzzy c-means clustering. In a fuzzy world, human beings have adapted to vagueness by not relying solely on quantitative measurement but by substantial use of qualitative information on a flexible basis. A question arises regarding the effective ways of representing fuzziness in qualitative information processing and reasoning. Fuzzy sets were first devised as ways of capturing the continua of membership grades conveyed by linguistic variables such as 'very liable' and 'possibly unsuitable'.

A useful method for encoding fuzziness with qualitative reasoning is by pairwise comparisons driven by inconsistency analysis (Saaty, 1978). This is particularly useful for evaluating alternative decisions against multiple criteria (Banai, 1993). But this approach is not discussed further here.

6.4.2 Analysis of Fuzziness

Fuzzy categorical data offer information on continua of spatial categories, and seem to suggest a self-contained system for coping with fuzzy classes, as has been discussed in the preceding section. However, the adoption of fuzzy approaches to information processing at the same time presents various issues, one of which is the non-trivial problem of how to characterise fuzziness in the first place. As fuzzy categories are non-crisp, it is possible to assert that fuzzily classified data will not be characterised in the context of objective ground truth. Moreover, every effort to define membership values for inherently fuzzy classes is bound to be approximate, however reasonable it may appear to be (Dubois and Prade, 1997). Thus, it may be more sensible to consider balancing fuzziness with crispness and its probabilistic languages so that fuzziness in a particular application context becomes less elusive and more interpretable.

Suppose a field domain D has been classified into continua of fuzzy classes, with individual locations mapped with fuzzy membership vectors. Fuzzily classified data are represented usefully as spatially varying fields of categorical continua, which provide informative visualisations of fuzziness. If data from heterogeneous sources are available to derive fuzzy classifications, it is possible to combine fuzzily classified data to arrive at augmented fuzzy data. Assume that two sources of data have been used to derive fuzzily classified maps with the same fuzzy-classification schemes denoted by U_k ($k=1,...,c$, c being the total number of fuzzy classes). For a location x_s in the field domain D, if the two sources of data lead to fuzzy data represented as $\mu1_k(x_s)$ and $\mu2_k(x_s)$, then fuzzy set operators AND and OR update fuzzy memberships by:

$$\mu1 \wedge 2_k(x_s) = \min(\mu1_k(x_s), \mu2_k(x_s)) \tag{6.20}$$

and:

$$\mu1 \vee 2_k(x_s) = \max(\mu1_k(x_s), \mu2_k(x_s)) \tag{6.21}$$

where the symbols $\mu1\wedge2$ and $\mu1\vee2$ stand for fuzzy set AND and OR operations respectively, and min and max are short forms for operations returning the minimum and maximum values of the fuzzy memberships in parentheses.

In addition to using entropy for measuring fuzziness, fuzzily classified data may be further analysed using a variety of techniques. Fuzzily classified data sets may be compared with respect to their goodness of approximation to the underlying vague classes. A useful measure is the distance measure used to assess the closeness of one set of fuzzy data to another, with the latter being taken as the reference, and believed to approximate the underlying fuzziness more closely. For a location x_s, a general distance measure between a fuzzy membership vector ($\mu_k(x_s)$, $k=1,...,c$) and its reference ($\mu0_k(x_s)$, $k=1,...,c$) is calculated as:

$$\text{fuzzy distance} = \left(\left|\mu_k(x_s) - \mu0_k(x_s)\right|^q + ... + \left|\mu_c(x_s) - \mu0_c(x_s)\right|^q\right)^{1/q} \tag{6.22}$$

where | | stands for absolute value, and positive real power q and its reciprocal $1/q$ define different metrics: $q=2$ for Euclidean distance and $q=1$ for Hamming distance.

Though distinctive on theoretical grounds, fuzzy sets may still be employed in combination with probability theory for handling real-world problems. Indeed, both bodies of theory are meant to deal with different aspects of uncertainty in complementary ways (Nguyen, 1997). It is in light of this argument that the notion of consistency may be exploited to indicate how closely fuzzily classified data are related to reference data that may be defined in the domain of probability, with both sets of data continuously valued in the range between 0 and 1.

Suppose a vegetated land unit is remotely sensed for evaluation of heavy-metal-induced stress, which may be reasonably represented as a continuum of fuzzy memberships. Assume that it is possible to obtain a data set describing the probabilistic distributions of heavy-metal pollutants, perhaps from using geostatistics. A heuristic observation suggests that a decrease in the fuzzy membership of vegetation stress tends to imply a decrease in the probability of underlying pollution. Let a fuzzy categorical variable for location x be defined by a fuzzy membership vector ($\mu_1(x_s),...,\mu_c(x_s)$), while probabilistic data are evaluated by ($p_1(x_s),...,p_c(x_s)$), where c stands for the same number of discretised classes. These two vectors refer to different semantics for the hypothetical example of metal pollution, the former pertaining to the graded status of vegetation stress and the latter to a set of thresholds for pollutants. The distance measure defined in Equation 6.20 may be used to quantify their closeness. Another useful measure is consistency, as formulated by Zadeh (1978). The degree of consistency between a fuzzy membership vector and its probabilistic counterpart is expressed by the dot product:

$$\text{consistency} = \mu_1(x_s)p_1(x_s) + \ldots + \mu_c(x_s)p_c(x_s) \tag{6.23}$$

which may be usefully interpreted as the cosine of the angle between the two unit vectors in c-dimensional space, with larger values indicating smaller angles, and so greater proximity.

A possible application of this consistency measure is to use proportions of component classes in a mapping unit (pixel) on the ground as reference data for fuzzy classification (Foody, 1995). However, it would be wrong to view mixtures of discrete land-cover types in individual pixels as a form of fuzziness in the realm of fuzzy-set theory.

The widely recognised superiority of fuzzy over crisp categorical maps is due to their information richness in respect of class continua. A disadvantage is the increased requirement for storage: in the case of c classes, there are c layers of raster data as opposed to one layer of vector data. While this extra cost for expanded storage may well be justified by the need for well-informed fuzziness, and the decreasing cost of expanded storage and extra processing, it is nevertheless useful to seek crispness within a fuzzy structure. This sensible exploitation of the crisp metaphor in fuzziness is more important than the quest for storage compactness, as the section below explains.

Given the longstanding controversy over arbitrariness in defining membership functions, it is problematic to perform accuracy assessment for fuzzy categorical data, although it may be possible to comment vaguely about fuzzy classification as reasonable or dubious. Any set of fuzzily classified data may provide a reasonably good approximation of the underlying fuzzy concepts or classes, if fuzziness is tuned to a specific purpose or a real-world setting. However, it is sensible to explore fuzziness by counterbalancing it with surrogate crispness resulting from defuzzification, so that parallel analyses of crisp and fuzzy sets may be carefully compared and applied. Such a balance between the simple representation and analysis of crispness and the information richness of fuzziness may be achieved by presenting fuzzy maps as unions of defuzzified areal classes and fuzzy boundaries, corresponding to subsets exceeding or falling below suitably chosen quantitative α-cut levels (Zadeh, 1968).

To obtain a defuzzification of fuzzily classified data, just as for probabilistic categorical data, pixels (locations) are assigned to the classes to which they have the maximum belonging, measured by specific class membership functions. For example, location x_s is assigned to class U_k via the condition:

$$u(x_s) = U_k \text{ if } \mu_k(x_s) = \mu_{\max}(x_s) = \max(\mu_1(x_s),\ldots,\mu_c(x_s)) \tag{6.24}$$

where class labels U_k form the classified data layers, with k ranging from 1 to c, and s from 1 to n. An illustration of defuzzification of a fuzzy four-class case is provided in Figure 6.16.

Such defuzzified categorical data are less useful than expected, since they are deprived of information on fuzziness. But a balanced solution may be obtained by profiting from the fuzzy membership values readily available in fuzzy maps. The maximum membership value $\mu_{\max}(x_s)$ of location x_s in the study domain, identified via Equation 6.24, is examined with reference to a pre-determined threshold α such that a location x_s is selected with the class label denoted by $u(x_s)$, if the maximum membership value is not less than α. A resulting α-cut set $U(\alpha)$, that is, a selectively defuzzified categorical map, is described concisely as:

$$U(\alpha) = \{u(x_s) | \mu_{\max}(x_s) \geq \alpha\}, k = 1,...,c, s = 1,...,n \qquad (6.25)$$

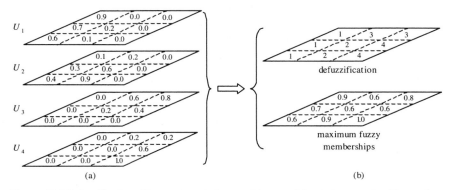

Figure 6.16 Defuzzification of fuzzy categorical maps: (a) maps of fuzzy class memberships; and (b) maps of defuzzification and corresponding maximum fuzzy memberships.

Clearly, the purity of the selected locations will increase with the threshold value. But this will be associated with a decrease in their number. Thus, the term 'pure pixel' is used only in a relative sense, and indicates those locations more similar to their respective class prototypes than others.

Once selectively defuzzified data are derived, it is possible to cross-tabulate the defuzzified and prototype class labels of pure locations to construct error matrices. The provision of both fuzzily classified data and their defuzzified subsets would enable an assessment of classification accuracy based on crisp measures such as the percentage correctly classified and the kappa coefficient of agreement, in addition to the valuable information on fuzziness already captured by fuzzy classification. Zhang and Foody (1998) provide empirical results suggesting the advantages of using fuzzy as well as crisp approaches to assessing fuzzy mapping of suburban land cover. However, it is important to recall that, in the strict sense, there is no substitute for an independent source of reference data, against which fuzzily classified data may be checked. In this sense, the accuracy assessment of fuzzy categorical data via selective defuzzification is, at best, a reasonable approximation.

The other aspect of selective defuzzification concerns the derivation of fuzzy boundaries, and serves to transfer the concept of fuzziness from semantic to spatial domains. Fuzzy boundaries are referred to as boundaries of non-zero width on fuzzy categorical maps, and have been discussed in different contexts (Wang and Hall, 1996; Brown, 1998). The remainder of this section will examine how fuzzy boundaries may be defined and derived from fuzzy categorical maps by using alternative criteria.

Clearly, fuzzy boundaries are complements of the locations selectively defuzzified by Equation 6.25. Specifically, a location x_s falls within the fuzzy boundary zones if its value of maximum membership $\mu_{\max}(x_s)$ is less than the threshold value α. Another criterion for defining fuzzy boundaries is the confusion index, which is evaluated as 1.0 minus the difference between the fuzzy membership values of the most likely and the next most likely classes. An example of fuzzy boundaries derived from simulated fuzzy maps was given by Burrough and Frank (1996). The assumption underlying such an index is that the greater the confusion index, the smaller the difference in fuzzy membership values between the most likely and the next most likely classes, and thus the more likely that the location defines a fuzzy boundary. The confusion index involves two fuzzy membership values for each location, and is thus more complex than the previous criterion. Again, a predefined threshold is employed such that a location defines a fuzzy boundary if its confusion index exceeds this threshold. Finally, the measure of entropy in Equation 6.17 may also be used for defining fuzzy boundaries, using all of the fuzzy membership values for each location. As entropy is a measure of disorder, it is logical to assert that boundaries usually occur where locations have high degrees of fuzziness, that is, high values of entropy.

It is worth stressing that fuzzy boundaries, as uncertain zones that border adjacent areal units, are more complex than the features expressed by epsilon error band models. Boundaries are not subjectively set *a priori*, but defined on the basis of suitable membership functions. The uncertain zones of epsilon error bands should be seen as post-processed or generalised boundaries of finite widths, conceptualised within the explicit object formalism of categorical mapping.

Empirical data were used to establish a hypothesis regarding the equivalence of the three alternative criteria in locating fuzzy boundaries from fuzzily classified maps (Zhang and Kirby, 1999). It was found that the correlation between maps generated using these criteria was statistically significant at a confidence level of 95%. This quantitative correlation analysis was supplemented by evaluating agreements between the fuzzy boundaries defined with these alternatives. For this, 30 levels of thresholds were chosen incrementally for the maximum membership values, but decreasingly for quantities based on the confusion index and entropy. It was required that the resulting number of locations (pixels) be equal between any two successive criterion values. In other words, the number of fuzzy boundary pixels located with a chosen criterion was expanded at a rate of about 3.33% for each increment in the threshold. Error matrices were then constructed to evaluate agreements in fuzzy boundary positions. It was confirmed that fuzzy boundaries were consistently defined within a confidence interval of 95%.

The distinctions among the three criteria described above for defining fuzzy boundaries can be examined further. The maximum membership values are straightforward to interpret and are closely associated with the conceptual gradation of class membership. The confusion index involves memberships belonging to the most and the next most likely classes. The interpretation of the confusion index is thus not as straightforward as for maximum fuzzy-membership values, suggesting a weaker relationship to conceptually graded fuzziness. Entropy appears as a measure of degree of fuzziness on its own. It requires a complex logarithmic calculation using all elements of fuzzy-membership vectors. As the selection of the base for logarithmic calculation is largely arbitrary, the interpretation of entropy is thus not straightforward. Therefore, a set of fuzzy-categorical maps can be represented compactly via a defuzzified class map and its corresponding map of maximum membership values with little loss of fuzziness. Clearly, a well-balanced representation of fuzzy categories is obtained by defuzzified discrete maps of classes accompanied by location-specific fuzzy-membership values, which can later be used to define fuzzy boundaries in practical applications.

6.4.3 Mapping Continua of Land Cover and Vegetation

In order to represent inherently fuzzy spatial categories such as gradation of land acidification and continuum of wildness, the preceding section has described fuzzy sets and their applications for spatial vagueness. A suitable fuzzy classifier may be tested and chosen for a particular application on the basis of validity and functionality.

Fuzzy sets hold conceptual advantages in dealing with vagueness in reasoning and decision-making, but the interpretation of fuzziness sometimes remains elusive. Indeed, the need to specify application-dependent membership functions renders fuzzy approaches to categorical mapping theoretically less appealing, and the arbitrariness in characterising fuzzy memberships is often singled out for criticism. A fair and empirical evaluation of fuzzy sets for spatial data handling is necessary so that their applications are well guided.

In Section 6.2, empirical results were obtained using probabilistic approaches to mapping probable occurrences of land cover in an Edinburgh suburb. Five broad classes were assumed to be discernible at the test site and thus crisply defined. This section will present another case study using the same Landsat TM data and training and testing data listed in Table 6.2. Fuzzy classification was carried out not only to demonstrate fuzzy classification but also to explore the potential for corroboration of fuzzy vegetation continua or spectrally graded land cover with probabilistically classified land cover. Since the set of land-cover types is assumed to be crisp, at least in relative terms, the fuzzy approach used here is not intended to challenge that assumption.

It is interesting to expose any (dis)similarity between fuzzy and probabilistic classification by examining how land-cover types are classified by the two distinctive methods. This is possible by using a supervised mode of fuzzy classification, in which fuzzy-membership functions are elicited from training data that support an approximate yet more explicit definition of land-cover memberships. Such a fuzzy classifier is built on histograms of spectral data.

Shown in Figure 6.17 is a hypothetical histogram indicating the distribution of the relative frequencies of certain measurements (e.g., digital numbers of spectral reflectance). Such histograms can be created from training data consisting of a number of pixels representative of their class prototypes, which are rarely uniformly distributed. An important feature for the fuzzy definition of class membership is the maximum relative frequency of a named class, relevant to a specific data source, as denoted by *fmax* in Figure 6.17.

Suppose a pixel x is to be fuzzily classified, and has a digital number dn, which is associated in the histogram with a relative frequency fx. It seems reasonable to take the ratio $fx/fmax$ as approximating the fuzzy membership of pixel x in class U_k, that is:

$$\mu_k(x) = fx / fmax \tag{6.26}$$

where fx and $fmax$ stand respectively for the relative frequency of value dn and the maximum relative frequency recorded for class U_k.

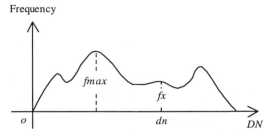

Figure 6.17 Histogram for defining fuzzy class memberships.

If there are several layers of fuzzy-classified maps, it is constructive to combine them in order to augment this information about fuzzy-class membership. The same combination of Landsat TM bands as in probabilistic classification was used to drive a fuzzy classification of land cover. By evidential reasoning, Landsat TM data were mapped into fuzzy-membership values. Results are reported in Figure 6.18, where graded continua of land cover based on spectral data are shown, (a) for grassland, (b) for built-up land, (c) for woodland, (d) for shrub, and (e) for water, respectively.

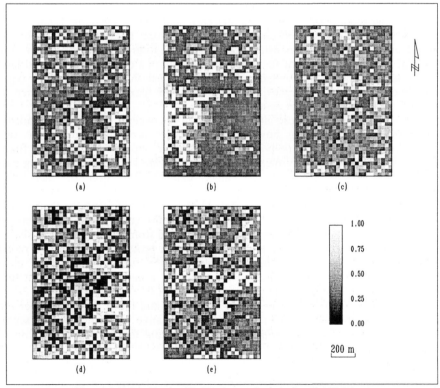

Figure 6.18 Fuzzily classified land cover from the Landsat TM data:
(a) grassland; (b) built-up land; (c) woodland; (d) shrub; and (e) water.

Fuzzily classified land-cover types appear strikingly varied over space. To obtain some numerical measures, the fuzzily classified map shown in Figure 6.18 was defuzzified and then checked against the test data in Table 6.3. This resulted in a PCC of 34.5%, significantly lower than that obtained with probabilistic classification of the same Landsat TM data. This may be accounted for by the imposed fuzzy definition of crisp classes of land cover, which is inappropriate for the problem at hand, whereas the probabilistic approaches described in Section 6.2 seem to model the occurrences of land-cover types well, in particular, by incorporating *a prior* probabilities. Thus, the distinction between probabilistic and fuzzy approaches to spatial classification is convincingly exposed by this real-data experimentation.

It is interesting to seek links between probabilistically classified land cover, which contains vegetated types, and fuzzily classified characteristics of vegetation. The dot product in Equation 6.23 is believed to suggest closeness between a set of fuzzy-categorical data and a related set of probabilistic maps. Thus, the dot product of the fuzzy and probabilistic memberships, which are displayed in Figures 6.18 and 6.5 respectively, was calculated to have a value of 0.31. Another quantity for closeness between fuzzy and probabilistic data is the Hamming distance, which is 1.11 for these data sets.

Fuzzy *c*-mean clustering is a classical method for creating fuzzy-categorical maps. By specifying the number of fuzzy classes desired, a clustering routine is typically run in an iterative mode to arrive at the globally minimised error defined in Equation 6.16. Using the Landsat TM image subset already introduced in Section 6.2, a fuzzy-clustering algorithm was programmed to drive fuzzy land classification based on spectral data or their derivatives. Some vegetation properties, such as vegetation index and vegetation quantity, are known to be indicative of land characterisation, and may be usefully calculated from a combination of the original Landsat TM image bands. These derived vegetation properties can further be fuzzily classified into a certain number of graded continua.

In this study, the normalised difference vegetation index (NDVI) was used to derive fuzzy membership values of individual pixels belonging to five grades of NDVI continua, since NDVI and other vegetation indices have been used to obtain land-cover classifications in previous studies (DeFries and Townshend, 1994; Campbell, 1996; Maselli *et al.*, 1998; Turner *et al.*, 1999). The ratio of the difference against the sum of Bands 4 and 3 was taken as the NDVI quantity, and a fuzzy five-means clustering was implemented. The process of clustering, with *m* set at 2.5, converged after 29 iterations. The results are shown in Figure 6.19, where five fuzzy classes of NDVI are shown, together with the centres formed during the process of clustering.

As shown in Figure 6.19, fuzzily clustered NDVI shows strong spatial variation. This provides an effective display of the graded continua of vegetation prevalence or abundance at the level of individual pixels. This set of fuzzy maps was compared to the set of probabilistically classified maps of land cover shown in Figure 6.5 using the dot product and Hamming distance. The resulting measures were 0.22 and 1.34 for the dot product and Hamming distances respectively.

Improvements were sought in the correlation between the fuzzily clustered NDVI and the probabilistic reference data. As the NDVI quantity indicates the abundance or vigour of vegetation at individual locations, it seems reasonable to assert that the graded NDVIs in the sequence 3, 2, 5, 4, and 1 may be more closely related to the land-cover types grassland, built-up land, woodland, shrub, and water. The resulting values of the dot product and Hamming distance were 0.24 and 1.30, which indicates only slightly closer correlation with probabilistically classified land cover. In general, these values consistently suggest that fuzzily clustered NDVI classification is even more different from the probabilistic reference data than fuzzily classified land cover.

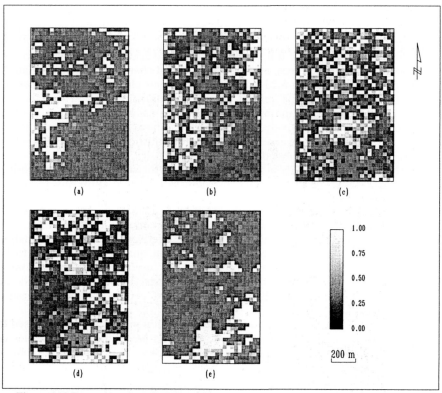

Figure 6.19 Fuzzy clustering of NDVI quantities based on the Landsat TM data with respect to five cluster centres: (a) 0.17; (b) 0.29; (c) 0.39; (d) 0.48; and (e) 0.58.

As in the case of probabilistic classification documented in Section 6.2.4, fuzzily classified maps can be defuzzified, along with the accompanying maps of fuzziness in the allocation of fuzzy classes. For this, the fuzzy maps shown in Figure 6.18 and 6.19 were defuzzified. The results are shown in Figure 6.20, where pairs (a)–(b) and (c)–(d) represent, respectively, maps of defuzzification and maximum fuzzy memberships created from fuzzily classified land cover and fuzzily clustered NDVI quantities based on the Landsat TM data.

Recall the methods outlined in Figure 6.16. The defuzzified maps shown in Figure 6.20 may be conceived of as containing the basic data for creating categorical maps accommodating fuzzy boundaries. The impression of spatially varied fuzzy memberships in (b) and (d) of Figure 6.20, however, implies that many inclusions of heterogeneity will be scattered within the fuzzy classes. This is typical of fuzzy classification based on spectral data, since fuzzy membership functions defined per-pixel will be imprinted with spectral heterogeneity picked up by remote sensors. Similar observations were made by Zhang and Stuart (2001), who experimented with approaches for extracting fuzzy boundaries from fuzzily classified images. Contextual approaches to post-processing per-pixel classification may be usefully extended to fuzzy classification (Gurney and Townshend, 1983), which help render spatially constrained fuzzy classification, with fuzziness purged from fuzzy class centres to boundaries, as is suggested in the geostatistical approaches to deriving the maps shown in Figure 6.11.

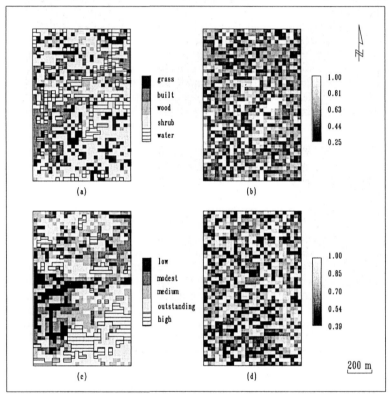

Figure 6.20 Maps of defuzzification and maximum fuzzy memberships from: (a)–(b) fuzzily classified land cover; and (c)–(d) fuzzily clustered NDVI, based on the Landsat TM data.

6.5 ROUGH ASPECTS OF CATEGORICAL MAPPING

6.5.1 Incompatibility of Information and Data Classes

Both probabilistic and fuzzy methods have been discussed previously in this chapter in an attempt to accommodate uncertainties in spatial classification. Probabilistic and fuzzy approaches to mapping continuous memberships are oriented to handle uncertainties due to categorical randomness and vagueness, respectively. Both have been widely tested and shown to outperform crisp or hard classifications, because of their adaptability to spatial variability in class occurrence and spatial continua of fuzziness.

Upon reflection, an implicit assumption often made in mapping crisp classes is that categorical truth exists and is unambiguously defined, at least in principle, and is retrievable upon inspection *in situ*, or through ground verification. Classification schemes for categorical variables are usually designed to be exhaustive and mutually exclusive, and conceptual or information classes are defined on this basis. It is often reasonable in a spatial classification to assume that data obtained by field surveys or from existing maps or aerial photographs are of relatively higher accuracy than remote-sensor data of coarser spatial resolution, and are taken as surrogate truth or reference pertaining to information classes.

It is useful to conceive of data-driven classification as transferring information classes from data spaces. This process is well known to be prone to misclassifications, since it is generally understood that data-driven classification rarely emulates human perception, and human interpretation of spatial categorical truth. On the other hand, the process of transfer by humans is frequently assumed to be of high fidelity, leading to an acceptance of human knowledge as providing the equivalents or prototypes of information classes, which are in turn used to benchmark a specific set of classified data. The accuracy tests used in this book have focused on comparing data-driven classification and reference data instead of truth, implying the possibility of biased reports of accuracy assessment, because ground verification of conceptually defined information classes may also be prone to error and is not necessarily replicable.

This challenge to the alleged superiority of human intelligence has a significant implication for accuracy assessment. Clearly, it would be simple-minded to denounce the potential of remotely sensed images for inadequate accuracy when reference data are themselves uncertain. Any disagreements between classified and reference data sets should not only be seen as indications of possible misclassification by classifiers, but also be taken as symptomatic of the uncertainties present in the chosen reference data. As reference data are usually under-sampled due to prohibitive cost, they should only be extrapolated over space with great care. Therefore, there is no guarantee that ground-observed or photo-interpreted reference data are more accurate than classifications based on remote-sensor data, and any superiority of the former over the latter should not be attributed solely to the spatial resolutions typically encountered in remote sensing.

For instance, a question arises about whether the classification of forests is directly associated with the identification and grouping of trees. It may also be doubtful whether intermingled grass and soil should be generalised into the dominating forest cover, or the heterogeneities recognised. Remote sensors typically capture all on the land surface that is spectrally visible within their instantaneous fields of view, which certainly may include trees, grass, and soil, but not exclusively. Therefore, a test of classified data against a reference would have implications beyond a simple count of disagreements, if the two sets of data could not agree on whether a distinctive group of trees should be retained or blended with others.

Spatial classification relies on careful selection of discriminating features. Using remote sensing techniques, multispectral reflectance data are typically used for land classification. Unless certain spatial categories have distinctive spectral signatures, data-driven classification is bound to be of limited accuracy. For forest/non-forest classification of cloud-screened European AVHRR data, Roy *et al.* (1997) divided their study area into 82 ecological and climatic strata. It was found that a combination of NDVI and surface temperature features produced maps of consistently higher accuracy than using either alone.

Early work by Robinove (1981) commented on the logic in classification and mapping of land in general terms, and warned against inconsistency or incompleteness in describing data-driven classification as a surrogate for information classes. Uncertainties in ground data were discussed by Fuller *et al.* (1998), who echoed the concern over discrepancies in class definition and interpretation. Intriguing scenarios like these reflect ubiquitous uncertainties in the perception and mapping of spatial categories.

With these considerations, it is useful to look elsewhere for formal expression and handling of incompatibility between information and data classes. Developments in rough sets are of great relevance, since they deal with the explicit analysis of incompatibility between information and data classes.

6.5.2 Indiscernibility, Rough Sets, and Rough Classification of Land Cover

Methods based on rough sets have been applied to an increasing extent in geography. Worboys (1998) explored imprecision in spatial data caused by finite resolution or granularity of spatial observations, using methods similar to rough sets but extended to include both spatial and semantic dimensions in developing models for geographical vagueness. Rough sets were explored further by Duckham *et al.* (2001), who considered an ontology of imperfection with practical implementation in retailing geography.

The theoretical thrust of rough sets is based on the assumption that information can be associated with elements in the universe of discourse (Pawlak, 1982, 1998). Subsets of elements characterised by the same information are indistinguishable, and thus form elementary sets or concepts that can be understood as elementary granules of knowledge about the real world.

Any union of elementary elements is a crisp set, and sets constructed otherwise are rough. Any set can be associated with two crisp sets called the lower and the upper approximation. The lower approximation of a set is the union of all elementary sets that are included in it, whereas the upper approximation of a set is the union of all elementary sets having non-empty intersection with it. In other words, the lower approximation of a set is the set of all elements that surely belong to it, while its upper approximation is the set of all elements that possibly belong to it.

The difference between the upper and the lower approximations of a set is its boundary region. Thus, a set is rough if it has a non-empty boundary region. Elements of the boundary region cannot be crisply classified by employing the available information, either to the set A or to its complement. Approximation of sets is the basic operation in rough sets and is used as the main tool to deal with uncertain data.

An example is provided in Figure 6.21, in which a polygon labelled U_k is approximated by granule elements of a regular raster. Suppose polygon U_k is a set consisting of an infinite number of points falling within its interior and boundary. Approximating U_k by raster cells of finite size leads to roughness in the data representing the original polygon. In Figure 6.21, the dark-shaded raster cell represents the lower approximation of U_k, since the cell is surely included in it; all of the raster cells except for the single white cell comprise the upper approximation of U_k, which over-estimates the area it covers. The difference between the upper and lower approximations is the border set, which is shown as lighter-grey-toned raster cells in Figure 6.21.

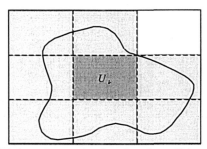

Figure 6.21 A rough approximation of a polygon U_k by raster cells.

Figure 6.21 has given a geometric interpretation of rough sets, and the remainder of this section is devoted to further description of roughness. Data about elements of interest are often presented in a table listing individual elements as rows that are described by columns of attributes. A special kind of table consisting of classification rules as columns of condition and decision is relevant to categorical mapping. For spatial classification based on spectral reflectance data, it is possible to regard relevant spectral bands of the sensor measurements as columns of a classification-rule table, while the decision column corresponds to the desired classification.

In the taxonomy of rough sets, any subset of attributes, denoted by AT, determines a binary relation on the universe of discourse, which will be called an indiscernibility or equivalence relation. With respect to a specific AT, all indiscernible classes are crisply defined and can be coded into subsets of c classes denoted by $\{E_1,...,E_c\}$. If two elements x_1 and x_2 are indiscernible from the perspective of AT, that is, if $AT(x_1)=AT(x_2)$, they are said to belong to the same class E_k determined by AT.

For instance, all pixels that have identical or similar spectral signatures are indiscernible in view of the available information, which consists of elementary sets or concepts that can be understood as elementary granules of knowledge about pixels, or rather about their equivalent ground locations. These elementary granules of knowledge may pertain to qualitative or quantitative information concerning these locations. Categories discernible with spectral data are known as data classes in remote sensing.

Consider a field domain D. Suppose equivalence classes are created from data AT and are coded as $E_{k'}$ with k' ranging between 1 and c'. The lower and upper approximation of a particular information class U_k, denoted by U_{k*} and U_k^*

respectively, are defined as:

$$U_{k*} = \left\{ E_{k'}(x) \subseteq U_k \,\middle|\, x \in D \right\} \tag{6.27}$$

and:

$$U_k^* = \left\{ E_{k'}(x) \cap U_k \neq \varnothing \,\middle|\, x \in D \right\} \tag{6.28}$$

where equivalence classes $E_{k'}$ are traversed from the set of data class labels $\{E_1,\ldots,E_{c'}\}$. From these, the boundary region of U_k can be defined as:

$$BN_k = U_k^* - U_{k*} \tag{6.29}$$

A rough set can be characterised numerically by the coefficient or accuracy of approximation:

$$\alpha_{AT} = |U_{k*}| \big/ |U_k^*| \tag{6.30}$$

and by the notion of a rough membership function:

$$RF_k = |U_k \cap E_{k'}| \big/ |E_{k'}| \text{ for } E_{k'}(x) \cap U_k \neq \varnothing \tag{6.31}$$

where the symbol $|.|$ denotes the cardinality of a set, and data classes k' range between 1 and c'.

Clearly, the accuracy of approximation by rough sets takes values between 0 and 1. For the example shown in Figure 6.21, its accuracy of approximation by raster cells is valued as the ratio of 1 against 8. A value of 1 for accuracy of approximation would indicate non-rough or precise approximation where there is a null boundary region, with lower and upper approximations coinciding. This would only be possible if a polygon were delineated by vectors enclosing pure points and lines. Furthermore, the rough membership function RF bears striking resemblance to the formulation of the conditional probability of information class U_k recovery given data classes $E_{k'}$ having non-empty intersection with U_k.

It has been discussed in Section 6.4.2 that fuzzy boundaries are spatial mappings of α-cut sets of fuzzy classes. In rough sets, the lower approximation is equivalent to the concept of cores in fuzzy sets, whereas the upper approximation relates to the notion of support of a fuzzy set. As ways for diminishing subjectivity in defining membership functions, rough sets seem to stand as some sort of compromise between objectivity and subjectivity for handling vague categorical data. A seemingly plausible way to proceed might be to concentrate on cores (relatively pure and certain) and borderline zones (uncertain) in classification, in pursuit of practical improvement over conventional fuzzy methods. This would be comparable to the categorisation of certain and uncertain zones in rough sets. But doing so will involve a large amount of subjectivity. In principle, rough sets are based on the manipulation of crisp rather than fuzzy sets. Thus, rough sets should not be seen as a special treatment of fuzzy sets but as a further technique offering potential for uncertainty handling.

It is worthwhile to explore the association between information and data classes in the context of rough sets. An illustration of the difference between what is assumedly the true classification, denoted as the desired information classification, and what is derived from available data is shown in (a) and (b) of Figure 6.22, respectively.

In Figure 6.22, the raster cells (pixels) are associated with integers, and are classified as shaded or blank. Suppose there is no error due to spatial resolution, contrary to that shown in Figure 6.21. The focus for discussion is placed on agreement between true and data-driven classification. Table 6.5 tabulates the association between information classes and data-driven classification for individual raster cells for the classification shown in Figure 6.22, where binary codes 1 and 0 of the classification correspond to whether a pixel is shaded or left blank. The final column of this special type of *AT* is titled compatibility, and will be discussed further.

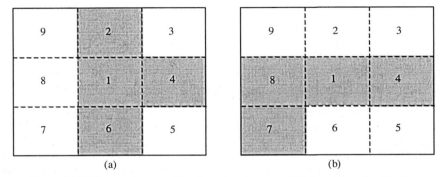

(a)   (b)

Figure 6.22 Difference between: (a) information classes; and (b) data-driven classification.

Table 6.5 Association between information and data-driven classification of binary codes.

Pixels	Data	Information	Compatibility
1	1	1	√
2	0	1	×
3	0	0	√
4	1	1	√
5	0	0	√
6	0	1	×
7	1	0	×
8	1	0	×
9	0	0	√

As indicated in Table 6.5, five out of nine pixels, {1,3,4,5,9}, are correctly classified, and are given ticks in the compatibility column. The correctly classified pixels are not readily separable from the perspective of rough sets, because the division of information class {1,2,4,6} by data classes {1,4,7,8} and {2,3,5,6,9} is far from clean.

However, for those incorrectly classified pixels that are marked with incompatibility (×) in Table 6.5, further differentiation is possible. Pixel set {2,6} is misclassified from 1s to 0s by the available data, while the set {7,8} is mis-labelled as 1s instead of 0s. If these two subsets of pixels are considered as distinctive classes and analysed alongside the correctly classified pixels {1,4} and {3,5,9}, class compatibility may be improved. Such a subdivision of classes can be facilitated by constructing an error matrix tabulating data-driven classification against assumed information class allocation.

A test was performed to experiment with the idea of rough classification. The same data set that has been shown in Table 6.3 was used. Analysis of initially classified data revealed that only 138 out of 168 training pixels were correctly classified, suggesting a PCC of 82.1%. This measure of PCC for training data seemed to have acted as the upper limit of classification accuracy for probababilistic classification using the different techniques in Section 6.2.4.

An extended set of classes entailing classifier indiscernibility was constructed by comparing the initially classified training pixels against their reference classification. The evidential approach to classification, as illustrated in Table 6.1, was applied under this extended framework of classes. It was found that all 168 training pixels except for one were correctly classified into their assumed information classes. With reference to the testing data, a PCC of 61.3% was obtained for evidential classification utilising information and data class compatibility, which is significantly higher than the 51.2% evaluated for the simple probabilistic classified data shown in Figure 6.7.

The formation, by rough sets, of class subsets that are discernible with given data is very suggestive of the evidential approaches to designing frames of discernment for reasoning about uncertainty in categorisation, as discussed in Section 6.2. In particular, the evidential constructs of belief and plausibility bear an impressive resemblance to lower and upper approximations by rough sets, and the interval of probability assignment in a particular class corresponds to the borderline in rough classification. When the existing evidence is not sufficient for classifying a location with reasonable accuracy into a discrete label, classifiers should be able to suspend classification. Moreover, accuracy should be evaluated by giving due tolerance to 'maybe' class labels. Classification accuracy may become inflated by relaxing restrictions on class agreement. In this sense, rough sets act in similar ways to fuzzy sets in allowing for softness in class memberships.

6.6 REFLECTION ON PROBABILISTIC, FUZZY, AND ROUGH METHODS

Spatial categorisation is a classic geographical problem. Categorical data are useful as a basis for deriving area estimates for spatially distributed categories of interest. Grid dot counting was a classic form of area estimation in forestry (Barrett and Philbrook, 1970). Developments in satellite remote-sensing technologies have been increasingly used to provide areal data (Bauer *et al.*, 1978; Card, 1982). Crapper (1980) examined the issue of errors in areal estimation from remote-sensor data. Later research coupled with GIS developments has led to enhanced techniques for spatial categorisation, and recognition of various aspects of uncertainty.

It has been shown in this chapter that the rationale for modelling spatial categories with discrete structures—polygons, or their raster counterparts—can be questioned in the context of realistic complexity. The probabilistic interpretation of class occurrences has improved the way categorical variables are represented and processed in a digital environment. Stochastic simulation has proven its versatility for modelling uncertainty in class occurrences, and the impacts of uncertainty on derivative data products such as areal summation using remotely sensed data.

The electromagnetic radiation data collected by remote sensors suffer scattering and absorption by gases and aerosols while travelling through the atmosphere from the Earth's surface to the sensors, which contribute to errors in reflectance data. In remote sensing of continuous quantities such as crop yields, it is generally understood that spectral reflectance data and their errors have direct bearing on model parameters. For example, in forecasting wheat production in various geographical areas using Landsat measurements and ancillary data on meteorology, growing season, and planting, irrigation, and fertilisation practices (MacDonald and Hall, 1980), the effects induced by spectral data errors are clearly worthy of investigation. However, atmospheric effects in reflectance data tend to have more complex impacts on the derivation of probabilistic maps than on continuous variables. For many applications involving classification and change detection, atmospheric correction is unnecessary as long as the training data and the data to be classified are commensurate. In other circumstances, corrections are required for registering multi-temporal data on the same radiometric scale in order to monitor the Earth's surface over time (Song *et al.*, 2001). Analytical modelling of radiometric and spectral inaccuracies, such as atmospheric effects, may be pursued, but stochastic simulation can be more usefully employed for modelling the sensitivity of probabilistic categorical mapping to errors as well as the probabilistic occurrence of spatial categories.

Probability is fundamentally related to frequency, but from a subjective viewpoint probabilities are states of mind and, hence, not uniquely quantifiable. In other words, the subjective interpretation of probability is often seen as the degree of belief of an idealised rational individual, and is not constrained to the frequentist's limited definition (Kyburg and Smokler, 1964). In evidential reasoning, it may be possible to define membership functions from histogram frequencies, or from the relative intensities of the frequencies of spectral measurements of an unknown pixel against a class prototype, as in the tests documented in this chapter. Thus, although subjective probabilities may be modelled mathematically, such modelling must be useful and meaningful for the applications at hand (Nguyen, 1997).

The pursuit of accountability in spatial categorisation extends to rough sets, which may have a greater claim to objectivity. But Koczkodaj *et al.* (1998) warn that the promise of objectivity may be misleading, since rough sets require an information table that often contains seemingly objective discrete data such as yes, no, low, or high. Such an information table is elicited from an expert's subjective judgments about raw data describing the real world. In spite of this distance between theory and pragmatism, this chapter showed the value of rough sets through an experiment. It was suggested that there is merit in exploring further the dissimilarity or inconsistency between image classification and assumed ground truth.

The notion of crisp sets is important for a variety of methods and applications including cluster analysis, Dempster-Shafer theory, and rough sets. In a discrete GIS application, it is important to realise that a spatial decision will depend on how the vague concepts, objects, or relations are transferred into crispness by applying certain thresholds (Fisher, 2000; Keefe, 2000). Another aspect of categorisation uncertainty originates from the fact that much of human reasoning and concept formation is fuzzy rather than crisp (Zadeh, 1997). Fuzzy sets have met with increased interest in the GIS and remote sensing communities. It is easy to gather arguments in favour of the strength of fuzzy sets as opposed to the weakness of crisp sets and theories built on crispness, but it is harder to clarify these two competing paradigms with tangible examples (Lark and Bolam, 1997).

Sections 6.2 and 6.3 discussed probabilistic approaches to spatial categorisation, which are seen to have good mathematical grounds, and to work effectively for modelling spatial correlated occurrences of crisply defined classes, in particular through the use of geostatistics. Section 6.4 provided both theoretical and empirical accounts of the flexibility and capability of fuzzy sets for mapping spatial continua. Clearly, probabilistic and fuzzy approaches are designed to cope with different but overlapping aspects of uncertainties, through randomness on the one hand and vagueness on the other. The remainder of this chapter is devoted to further clarification of probability and fuzzy sets.

A fuzzy set is a class of elements with a continuum of grades of membership (Zadeh, 1965). In geographical terms, a fuzzy class is characterised by a membership function that assigns to each location a grade of membership ranging between zero and one. It may be useful to consider some hypothetical examples, such as the assessment of vegetation density. If the location of an individual plant is known, and if it is accepted that vegetation cover is determined by the vertical projection of plant outlines onto the soil, one may feel close to an objective definition of plant density. Nevertheless, there is still a question of what to measure in order to recover the required quantities of vegetation density, because the concept is vaguely defined. But if vegetation density is fuzzily mapped into several graded continua, and if this is all that is needed for the application, then mapping vegetation density will cease to be a concern for field scientists or data analysts wondering about what precise measurements to make and where to locate samples. The relevance and usefulness of fuzzy sets for spatial problem-solving are obvious in such examples.

A similar situation occurs in measuring slope gradients. Apart from basic ambiguity in the size of the area over which to measure terrain slope, measurement to a precision of 1% is often not easy particularly over rugged terrain (Giles and Franklin, 1998). Devising more accurate ways of collecting data, that is, taking more measurements, or mapping in greater detail, is costly and unnecessary if one recognises the flexible ideas of class overlap and partial membership that are provided in the realm of fuzzy-set theory. Likewise, even if quantitative data are available, it would be hard to imagine how troublesome spatial decision-making might become for daily life, if people were forced to carry out precise measurement of all relevant properties.

Consider yet another example of measuring variables in the field. Standards and work manuals based on idealised requirements for class memberships do not always find a perfect match with reality, and humans are likely to make subjective decisions in assigning discrete class memberships, whether or not they are aware of the full complexity of reality. To be more objective, one may wish to relax the requirement of a strict match between each individual case and one of the defined alternatives, so that the deviation of a particular thing from an ideal model is taken to indicate categorical fuzziness. Fuzzy sets are clearly useful therein for capturing variation and grades of belonging to categories.

The conceptual advantages of fuzzy sets and methods are important in land evaluation, where combinations of variable values are needed to define required quality and suitability classes through logical operators. Class membership in discrete grades of suitability is conventionally defined by specifying the ranges of values of a certain number of key variables that an individual land unit must meet.

For example, one may assert that erosion hazard is severe if slope $\geq 10\%$ AND vegetation cover $< 25\%$. However, when a unit has values that are just outside the class boundaries—for instance, the slope is 9% and vegetation cover is 24%—crisp logic would exclude severe erosion. But most experts would suspect that erosion is still possible (Burrough and Frank, 1996). The superiority of fuzzy over crisp logic is apparent. Similarly, to remedy the deficiencies of crisp logic such as unrealistically sharp class boundaries and spatial independence of uncertainties across the map layers, fuzzy sets have been used for handling uncertainties in map overlay and have been found to be less sensitive to data errors (Burrough and McDonnell, 1998).

Fuzzy concepts even appear in statistical analysis and are handled by probability distributions. For instance, in testing the independence in a contingency table, the null hypothesis is rejected if the Chi-square statistic exceeds a critical value decided by the probability of error (Smith, 1961). But that probability is chosen subjectively, although subjective probability and fuzzy sets should not be confused due to their different theoretical underpinnings.

The versatility of fuzzy approaches has been demonstrated in a variety of scenarios, and fuzziness seems to be very close to the essence of human intelligence. Many realistic settings are naturally adaptable to or may benefit from fuzzy information processing. However, fuzzy sets should not be generalised to the point where inherently non-fuzzy things are treated on a fuzzy basis, when probability theory would be more suitable.

Mixed pixels or mixels, which contain mixtures of ground cover types, are problematic for discrete classification, and are typical of remotely sensed images when spatial resolution is too coarse to resolve the desired ground detail. Mixed pixels may also be related to multi-scale data with respect to spatial and categorical resolution (Franklin and Woodcock, 1997). Mixels have frequently been discussed in a fuzzy context, and an impressive amount of research results has accumulated. This is perhaps because the word fuzzy seems to make intuitive sense for spatial categories that are either conceptually vague, or appear fuzzy to the eye or to remote sensors. Mixed pixels have been discussed under the general label of fuzziness, although some of their components may well be treated on a probabilistic basis. It is nevertheless essential to recognise that fuzzy classes will not cease to be fuzzy at finer spatial resolution, while, in contrast, crisp classes will be resolved given fine enough spatial resolution. This is one of the defining characteristics that distinguish fuzzy-set theory from probability theory.

Fuzzy classes are sometimes referred to as classes having non-sharp boundaries. In geographical space, fuzzy classes may be considered as classes with indeterminate boundaries (Burrough and Frank, 1996). But caution is needed in this context. Probabilistic classification leads to a set of categorical maps that depict spatially varied probabilities of class occurrences at individual locations. Class boundaries of non-zero widths may also be elicited from probabilistic maps using maximum membership values as a criterion for thresholding (Zhang and Stuart, 2001). Similar representations and boundary derivations of fuzzily and probabilistically classified maps should not mask the difference in their underlying definitions of classes. Defuzzification with fuzzy boundaries does not make fuzzy concepts themselves non-fuzzy, while finite boundaries separating probabilistically classified classes do not deny the crispness of the underlying classes.

Perhaps the most challenging issue for fuzzy sets concerns spatial dependence in dealing with uncertain classification. In probabilistic approaches, spatial dependence can be accommodated so that fields of categories are correlated in the spatial domain or across variables, and where locations are categorised in ways that are consistent with certain stochastic distributions. Categorical variables, as important components in GIS and environmental modelling, are inherently difficult subjects for analytic modelling. But stochastic simulation is viable for modelling the process of spatial categorisation. It is hard to argue that fuzzy set theory provides any equivalents to joint and conditional probabilities, although probabilities of fuzzy classes can be defined in a way similar to that of Zadeh (1968). Therefore, it would be difficult to accommodate the effects of spatial dependence upon areal estimation from fuzzy classified maps, although fuzzy graded memberships at individual locations can be analysed within a geostatistical framework (Pawlowsky *et al.*, 1995).

A warning is given below against unwary use of probabilistic approaches to fuzzy sets. To determine a membership function for a fuzzy class, it may be possible to conduct a statistical survey and to take the proportion of people who asserted or classified a spatial entity or category as qualified for that class. There is clearly a strong flavour of frequentist probability in such derivations of fuzzy memberships. Given the distinction between the theoretical realms of probability and fuzzy sets, such a method of using probabilistic protocols for fuzziness is a sensible endeavour. But this is only one way to get an approximate functional model for a fuzzy concept or class, and does not address the inherent vagueness of fuzzy sets.

The overall sense is that theories of probability and fuzzy sets address aspects of uncertainty that are often disjointed. Probability theory is well established and widely accepted in decision-making, as well as in scientific experimentation and analysis. The estimation of uncertainty in terms of probability is the most conservative but the most reliable and mathematically sound. On the other hand, fuzzy-set theory is a conceptually flexible way of expressing and representing vagueness, but relies heavily on an assumed membership function to characterise a concept or class having inherent fuzziness. A further issue lies in combining fuzzy data, though there have been impressive developments in fuzzy-set theory and possibility theory (Zadeh, 1978; Viertl, 1997).

Although it may not be so hard to differentiate one from the other theoretically, it is often difficult to see a clear difference between them in practical terms. Instead, both theories should be seen as complementary, since the real world is full of multiple uncertainties, and a mechanical assignment of appropriate approaches is rarely possible or sensible. For example, the fuzzy classification of land-cover-related quantities demonstrated in the previous section was suggestive of the underlying crisp classes of land cover shown in Section 6.2, suggesting the existence of varying degrees of association between probabilistic and fuzzy interpretations of land cover. In view of this, there is great potential for the collaborative use of probabilistic and fuzzy categorical data, as explored in Section 6.4.3 by examining two alternative continuous methods for classification. Developments along this direction will be significant for geographical applications of probability theory, fuzzy sets, and rough sets. For instance, imprecision in probabilistic evaluation can be accommodated using fuzzy logic while accuracy in fuzzy memberships may be assessed with the notion of probabilistic distributions.

Spatial classification of such phenomena as suburban land cover implies a hierarchy of source data of varied accuracies. There seems to be conspicuous merit in maximising information potential by conflating data of varied uncertainty in an uncertainty-aware and sensible way. But any conflation effort has to confront many conceptual (semantic), spatial, and measurement issues associated with heterogeneous data. It is hoped that this chapter has built a working framework for exploring uncertainties in spatial and semantic hierarchies.

CHAPTER 7

Uncertainty in Objects

7.1 INTRODUCTION

As models of geographical abstraction, fields are known to possess functional advantages for the analysis of spatially distributed phenomena. The effectiveness and efficiency of fields for geographical modelling and data handling are widely appreciated and amply demonstrated in geographical information science and the geographical information technology industry. The main strength of fields seems to be particularly conspicuous for handling spatial uncertainties, as discussed in Chapters 5 and 6, where uncertainties in variables of both continuous and discrete types were effectively described and modelled through the exploration of spatial variability and dependence. It was also confirmed that field-based representational models and computational methods contribute substantially to a conceptual shift from deterministic description to stochastic modelling.

However, many geographical data exist more naturally as objects, which occupy an otherwise empty geographical space. In the context of urban planning and management, for example, the representation of complex urban landscapes requires the frequent use of well-defined and precisely valued objects for capturing entities such as buildings, road networks, and utilities. Political, administrative, and property boundaries are discretely defined as well, with positions and attributes that require precise measurement and evaluation. Even in the domain of fields, objects are employed for representing realisations of stochastic field variables or for facilitating geographical categorisation. Moreover, as was explained in Chapter 6, fuzzy classification can lead to a flexible derivation of fuzzy boundaries by ordinal scaling of fuzzy-categorical fields, and these may be conceptualised as line objects of finite width. Despite these advantages, there has been little explicit investigation of objects with respect to uncertainty. This chapter extends the stochastic perspectives embedded in the realm of fields into the domain of discrete objects, in an exploratory analysis of object uncertainty.

Objects may be represented as rasters of regularly shaped cells, or as vectors of discrete and irregularly shaped points, lines, and areas. The choice of suitable data structures often relies on the specific techniques used in data acquisition. Land surveying equipped with total stations may be used for accurate positioning, while field surveys are often the source of observations and measurements of such object attributes as building height and road surface material. Photogrammetric data are increasingly incorporated into the mainstream of mapping and management practice at a range of scales, although map digitising is still an important source of geographical data.

Many types of geographical data are naturally structured in vector form, by encoding the identities and spatial extents of real entities or abstracted objects, and possibly with relationships among them being computed or recorded. On the other hand, computerised data acquisition based on techniques such as digital photogrammetry, remote-sensing image analysis, and map scanning involves the use of raster structures. No matter what structures are used, objects are defined with attributes in addition to positions, enriching object semantics.

The position of an object is described by a set of coordinates, while attributes represent the qualitative and quantitative facts that are meaningful in specific problem domains. Position is implicitly represented in a raster format, while coordinate tuples defining position are explicitly recorded in a vector format, implying the possibility of numerical coordinates of high yet often spurious precision. Problems of spurious precision in positional data were initially discussed and investigated experimentally by Goodchild (1978), and they will be further addressed in connection with the accuracy inherent in object identification and positioning later in this chapter.

An interesting characteristic of objects is the separability of position and attributes for well-defined and unambiguously extracted objects. Clearly, accuracy in census-tract geometry has no bearing on the reliability of household inventory data, which may be affected by non-geometric factors including questionnaire completeness, correctness, and timeliness. As attributes are likely to be attached to specific applications, and as expert knowledge in a problem domain determines what attributes are relevant and how they are going to be used, the discussion of spatial objects often places more emphasis on positioning and positional data. For similar reasons, greater emphasis in error handling for spatial objects is placed on their positional dimension, unless there are gross errors in the observation and measurement of object attributes.

Positional data are subject to gross, systematic, and random measurement errors. Gross errors may be identified by inspection or by repeated measurement. Systematic and random errors are traditionally analysed through well-established statistical approaches. For example, surveying adjustment is a conventionally used tool for error analysis in position fixing by surveyors, and can estimate standard errors in X and Y directions and covariances between X and Y, in addition to calculating the most probable positions for individual points (Tyler, 1987; Cooper and Cross, 1988).

However, errors involved in positioning fixing are examined to a very limited extent, and are largely restricted to geometric primitives. For instance, in line plots produced by photogrammetric digitising, only simple error estimates such as RMSEs in (X,Y) coordinates are reported, providing global but not spatially explicit descriptions of positional errors, and precluding any appreciation and exploration of spatial variability in positional errors. The discussion of absolute and relative errors, though useful, tends to be confined to a conceptual level: absolute errors refer to individual points in isolation, while relative errors are associated with pairs of points. Certain points measured from maps or photographs may be treated independently in error analysis as if they were isolated individuals. But many linear entities or lines enclosing area objects are often digitised in stream mode, such that deviations from true positions are likely to be spatially correlated. Therefore, a more comprehensive treatment of errors in objects is needed for well-informed spatial decision-making in a GIS environment, in which points, lines and areas should be interpreted not only in geometrical terms, but also from the perspectives of spatial variability and correlation.

An important question concerns the uncertainty surrounding object identification, as there would be no point in discussing positional and attribute uncertainty without the ability to identify the objects reliably in the first place. It is assumed that objects are identified and delineated by careful visual interpretation and manual tracking of individual entities from ortho-images or maps, suggesting negligible uncertainty in the process of object identification and delineation. However, to develop a systematic strategy for uncertainty handling in the domain of objects, the issue of object identification should be examined further, through a holistic view of the processes of conceptual abstraction and physical extraction of objects. Such a view would also help to establish the underlying links between fields and objects. The following section will discuss the issue of object identification or extraction from images, and its implications for modelling positional error.

Positional errors are already a topic of some familiarity to geographical data communities, with error descriptors such as error ellipses and epsilon error bands being widely discussed and used (Dutton, 1992; Goodchild and Hunter, 1997). Error ellipses are most often associated with point features, and epsilon error bands with line features, and the latter are more commonly ascribed to the whole data set than to individual features.

As noted in previous chapters, the modelling of errors is more useful and important than their description, since it can lead to improved understanding of error propagation and the impacts of errors on applications. Consider the assessment of accuracy in a vector map, and derivatives such as estimates of polygon perimeters and areas. It may seem a simple matter to estimate the positional accuracy of point data digitised from aerial photographs by checking the measured coordinates of a few well-defined points such as outstanding landmarks against their assumed true coordinates, by using RMSEs for planimetric (X,Y) data (ASPRS, 1989). However, the task of evaluating accuracy becomes less straightforward when dealing with lines and polygons. It is common practice to overlay the test data set on an assumed reference data set to obtain estimates of digitising accuracy descriptors, such as epsilon error band width (Edwards and Lowell, 1996). But such error descriptors may not be used as error models to predict the accuracy of derivative data products such as line lengths and polygon areas by means of variance propagation, unless homogeneity and spatial independence of positional errors among neighbouring points are arbitrarily assumed, and such assumptions are almost always invalid. A feasible way around this problem would be to simulate alternative versions of object data sets in order to accommodate errors in these data sets and to assess their consequences in specific geo-processing operations such as map overlay.

The process of stochastic modelling enables a useful exploration of spatial variability and correlation of errors in discrete objects. This chapter takes positional errors as its major concern, since they accumulate in various forms in high-level objects. The results from an experiment using photogrammetric data will test the effectiveness of the proposed approach for modelling positional errors. The simulation approach will also be examined with respect to its advantages as opposed to other methods where due considerations are not given to the spatial correlations intrinsic to positional errors. Stochastic-simulation-based modelling of uncertain vector data via raster structures will be shown to be a valuable extension and contribution of geostatistical approaches to the integrated handling of errors in heterogeneous spatial data.

7.2 DESCRIBING UNCERTAIN OBJECTS

As mentioned in the introduction, an issue of fundamental importance for any discussion on uncertainty in objects concerns their identification and extraction, in addition to delineation and measurement. From time to time this issue has been overlooked or ignored, even by trained and conscientious personnel, and questionable assumptions have been made that objects are somehow unambiguously identified, accurately located, and attributed to any precision required. Realistic strategies for handling uncertainty in objects must avoid such simplified views of a geographical reality that is highly complex. Thus, it is worthwhile to challenge such unreasonable assumptions in object abstraction, and to find out how such an inquiry may provide a more unified view of field and object uncertainties.

Uncertainty mixes objectivity and subjectivity in different proportions depending on the specific methods employed for extracting objects from data sources. For objects extracted manually *in situ* or from images or maps, it is generally believed that data analysts perform careful identification of objects with a high level of accuracy. Objects extracted in such ways are believed to be of good quality, since any uncertainties associated with object identification occurred at an early stage in the whole information process. This assumption of certainty in object identification is not absolutely safe, and there are often few clues regarding how the objects were identified and extracted, or about the uncertainties in the process. Moreover, the human dimensions of uncertainty are by no means easy to quantify.

In a computerised data production context, on the other hand, objects such as houses and roads are identified from data sources such as digitised images or scanned maps, by referring to specific criteria or rules encoded as computer algorithms (Lillesand and Kiefer, 1994). Clearly, the fact that objects are elicited via programmable procedures indicates that uncertainty will be more tractable and suited to analysis than if the objects were the result of human-based data processing, which is dominated by visual and manual arbitration.

For example, sophisticated algorithms may be developed to extract objects of interest from geometrically rectified digital orthoimages, which may have been created from rigorously reconstituted stereo-models or by referring to a digital elevation model covering the same area. Such a process for object identification and extraction from images can be adapted to accommodate the occurrence of uncertainty specific to the particular data sets and algorithms involved. Similarly, meaningful points, lines, and areas may be retrieved from scanned maps, where case-dependent uncertainty can be addressed. This section will construct a framework for discussing uncertain aspects of object extraction from raster images and map sources, before proposing a general methodology for describing positional errors in vector data.

7.2.1 Uncertain Objects Extracted from Raster Data

Many real-world entities are well defined and lend themselves to object-based representation and manipulation. However, it is unwise to encode spatial objects as discrete points, lines, and areas without first thinking about the rationale underlying the abstraction of objects, and thus losing clues that may be important in understanding the uncertainties involved in object discretisation. The process of object abstraction and extraction has to be addressed directly if information is to be obtained about inaccuracy in spatial objects.

Raster structures, as a type of digital data representation, are employed for spatial data acquisition by remote sensing and map scanning, and are also recommended for geographical analysis as the basis of functions such as map overlay. Thus, it is useful to discuss uncertainty in object extraction with reference to raster data of finite size. An example is shown in Figure 7.1.

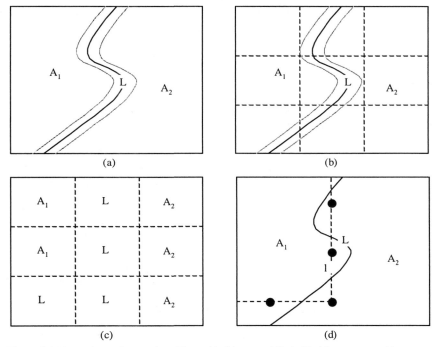

Figure 7.1 Abstracting and extracting objects: (a) objects modelled; (b) objects measured by a raster structure; (c) objects extracted and represented by raster cells; and (d) raster boundary approximation.

Illustrated in Figure 7.1(a) is a curved linear entity L, possibly of finite width as hinted by the grey-toned curves, which separates two neighbouring areas A_1 and A_2. Suppose these three objects are to be extracted by a raster-based system, as indicated in Figure 7.1(b), in which raster cells separated by dashed lines are superimposed onto the underlying objects. As a result of using the raster structure, the precise boundary detail is lost, along with any locational reference to points or lines of vanishingly small sizes or width relative to the raster cells. It is thus only possible to refer the object as contiguous groups of raster cells, as shown in Figure 7.1(c).

Positional data measured with a raster system are accurate to a degree that is bounded by the size of the raster cells (Müller, 1977; Theobald, 2000). To derive positional data for a line object such as L, it is possible to follow the centres of contiguous cells labelled as L in Figure 7.1(c), and to code this chain as an approximation to the position of the underlying object. An example is shown in Figure 7.1(d), where dots highlight the centres of raster cells within which a line object or edge is believed to be located. In Figure 7.1(d), the true and measured positional data are clearly differentiated with solid and dashed lines, respectively.

What is suggested by Figure 7.1 is that raster-based object abstraction and extraction become heavily tilted towards the semantic domain of objects, with positional data relegated to an implicit domain. For the extraction of line objects, it is possible to take the strength of boundary existence, evaluated according to the rule employed for edge detection in individual raster cells, as probabilistic data for the occurrence of object L. Such data are illustrated in Figure 7.2(a) for object L.

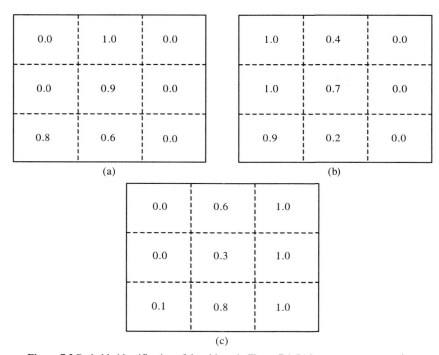

Figure 7.2 Probable identification of the objects in Figure 7.1 (b) in a raster representation: (a) line L; (b) area A_1; and (c) area A_2.

For area objects, some modification is needed. Consider an object A_1 shown in Figure 7.1(b). It is possible to examine the areal proportion it occupies in individual raster cells. For a raster cell to be labelled as area A_1, for example, it has to possess a dominant areal extent in light of the majority criterion that is generally implemented in raster data processing. The proportional areal extents of object A_1 in individual raster cells may be interpreted as probabilities that A_1 occurs, conditional to the measurement system and data available. An example for such probabilistic data is shown in Figure 7.2(b) for object A_1. Similar probabilistic data are shown in Figure 7.2(c) for object A_2.

Trinder (1989) discussed a model of positional accuracy in locating features from digital images. Aumann *et al.* (1991) described a technique for the computerised derivation of skeleton lines from scanned contours, another example of a probabilistic approach to the evaluation of uncertainty in derived lines. The process of image segmentation for the extraction of area objects can also be extended to include an extra but worthwhile evaluation of the uncertainty in extracted groups of contiguous pixels.

Tests were carried out to examine the uncertainty of object extraction from raster data, using subsets of the ortho-images shown in Figures 5.6 and 5.19(a), corresponding to the Edinburgh and Wuhan sites, respectively. Images of grey tones were transformed into the surfaces of gradient shown in Figure 7.3, where darkness represents the strength of detected boundaries.

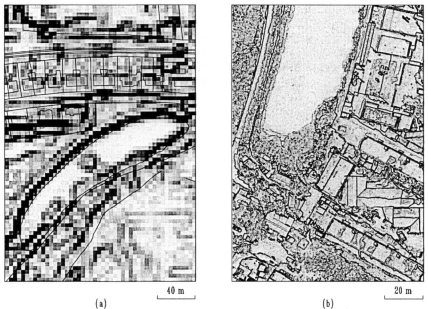

40 m

(a)   (b)

20 m

Figure 7.3 Boundary detection using gradient images based on:
(a) Edinburgh; and (b) Wuhan data sets.

Figure 7.3(a) concentrates on the loch to the north of Blackford Hill, while Figure 7.3(b) is located at the southern tip of the lake bordering the main campus of Wuhan University. As indicated in Figure 7.3(a), the coarse raster cells of about 4 m hardly resolve the fine detail of the superimposed vectors, although some boundary segments, notably the feather-like loch, are still detected. On the other hand, urban and suburban entities should be captured well when fine-resolution data are used. As shown in Figure 7.3(b), building geometries are identified with relative accuracy from the Wuhan sub-image at 0.2 m resolution, in spite of the persistent issue of determining spatially varying thresholds for adaptive boundary detection.

With sufficiently fine raster resolution and robust boundary detection, it should be possible to detect and extract many spatial entities with accuracy, and to audit the results with respect to image quality and algorithmic specifications. But computation, no matter how powerful, is no substitute for the human intelligence that is employed in much pattern recognition, and which underlies the vector overlay shown in Figure 7.3(a). It is seldom easy to balance the cost of human expertise and labour with the inaccuracies that are the inevitable result of data automation.

7.2.2 Vector-Based Sampling of Uncertain Objects

This section considers the acquisition of discrete objects by visual interpretation and manual delineation. An extracted (measured) object is different from the corresponding truth due to inaccuracy in object identification and positioning, as described above. Shown in Figure 7.4(a) is a vector data structure employed for approximating the same true objects indicated initially in Figure 7.1(a), where two areas A_1 and A_2 are separated by the true boundary line L. Curve l, the measured version of line L, which demarcates the measured footprints for areas A_1 and A_2, is shown as a dashed line against the original curve L shown as a solid line in Figure 7.4(a).

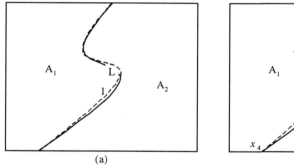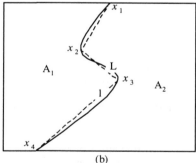

(a)   (b)

Figure 7.4 Measuring a true curve L with an approximation l: (a) as a curve; and (b) as a polyline.

In most discrete representations, real-world line objects are sampled by polylines that link up ordered vertices with straight line segments. If the real lines are truly curved, as shown in Figure 7.1(a), a polyline representation will be an approximation, and such differences between polylines and the original curves form part of the uncertainty in modelling objects. As shown in Figure 7.4(b), the true boundary line L, which is curvilinear, is sampled by selecting four vertices x_1–x_4 to result in the measured polyline l that consists of line segments linking up the four points. Since they are represented as vectors, the measured areas A_1 and A_2 are often called polygons.

This process of line sampling imposes an upper limit on the accuracy of the extracted objects, and its impacts should be borne in mind when performing error analysis. In other words, errors in objects are usually discussed with respect to sampled objects, which have already suffered the approximation and generalisation imposed by line sampling. Thus, any error models for lines should be able to resample the curved objects; otherwise, handling of uncertain objects using polylines is misleading.

Points x_1–x_4 marked in Figure 7.4(b) are sampled from the true line, and their measured positions are intended to coincide with the underlying true points. Thus, the sampled points are assumed to comprise part of the true positional description for line L. Now that this sampling process has been described, it is possible to move on to describing positional uncertainty. An example is given in Figure 7.5(a), where known vertices x_1–x_4 are measured, in one instance, to $x_{1'}$ –$x_{4'}$, respectively.

Positional errors, also known as displacements or distortions, are understood as the differences between the measured and the assumed true coordinates. Positional error vectors are shown in Figure 7.5(a), and the directions and magnitudes of point-specific errors are illustrated. In Figure 7.5(a), error vector $\varepsilon(x_r)$ at point x_r is taken as a general notion of positional error.

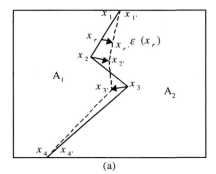 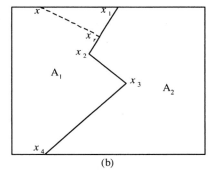

| (a) | (b) |

Figure 7.5 Positional errors: (a) error vectors with line vertices in true $(x_1$–$x_4)$ and measured $(x_{1'}$ –$x_{4'})$ position; and (b) a profile drawn from x_r, perpendicular to line segment x_1-x_2.

For convenience of visualisation, consider the line segment linking points x_1 and x_2 shown in Figure 7.5(b). It is possible to draw a line from x_r that is perpendicular to line segment x_1–x_2 and that intersects A_1's boundary at point x_0. A hypothetical probability density function (pdf) for positional error projected onto the direction of line segment x_r–x_0 is shown in Figure 7.6(a). Detailed discussion of positional error models will be the subject of the next section.

The correctness of identification can be represented by an indicator function $I(x;A_1)$, which is valued at one for points lying precisely on the dashed line between x_r and x_0 inclusive, zero otherwise, as shown in Figure 7.6(b). This indicator representation is based on the implicit assumption of accurate identification of discrete objects, such as area A_1 along the profile x_r to x_0, as crisp sets. The combined effect of the positional error modelled with pdf($\varepsilon(x)$) and the object identification of area A_1 along the profile is exhibited by Figure 7.6(c), where convolution of both results is the probability of point x lying in area A_1. A mathematical proof assuming smooth and small distortion was provided by Kiiveri *et al.* (2001). Clearly, this probabilistic evaluation of location with respect to a certain object leads to non-sharp boundaries for the object in question, as is indicated by the smooth curves localised at points x_r and x_0 shown in Figure 7.6(c).

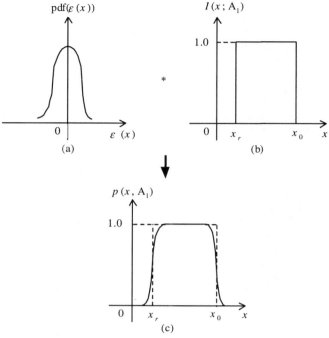

Figure 7.6 Object uncertainty: (a) a probability density function for positional error:
(b) object identification; and (c) combined effects.

It is clear that uncertainty in objects is not simply a matter of attribute and positional errors. Uncertainty in initial object extraction is more fundamental. In a raster-based system for object measurement and representation, uncertainty in objects goes directly to the issue of how accurately potential objects are extracted from discretised fields of imaged objects, and with what criteria. While raster cell size may provide a useful estimate of the upper bound of positional accuracy in extracted objects, there is no direct link between it and the location-specific positional errors that can be decomposed or singled out from the underlying potential fields of object occurrences.

Vector-based data structures may enable a discussion of positional errors at discrete points and lines. This is because the blurred boundaries shown in Figure 7.6 may be sharpened by decomposition of the convolution process, if the underlying objects are truly accurately defined. However, the prior process of sampling inherently curved objects by polylines already suffers from approximation, and is one source of the object uncertainties that affect subsequent spatial analysis and problem-solving. This is a problem that must not be overlooked; discretely sampled objects cannot be addressed simply as ensembles of individual points.

7.3 MODELLING POSITIONAL ERRORS

7.3.1 Analytical Models for Positional Errors

Illustrations provided in Figures 7.5 and 7.6 have set up a context for discussing positional errors in detail. A useful representation for an infinity of points x_r which lie on the line segment bounded by two end-points x_1 and x_2, is by linear combination of the form shown already in Equation 2.34 as $(1-r)x_1 + rx_2$ where scalar r represents the ratio of distance between x_r and x_1 to that between x_2 and x_1. Such a linear combination of vector points in representing lines holds advantages for computer cartography (Meek and Walton, 1990), a point that will be clearly demonstrated in the following text. So, the error vector at point x_r is a linear combination of the error vectors at the two end points:

$$\varepsilon(x_r) = x_{r'} - x_r = (1-r)\varepsilon(x_1) + r\varepsilon(x_2) \tag{7.1}$$

where error vectors are the differences between measured and known positions of the underlying points, and scalar r has the same interpretation as in Equation 2.34.

From Equation 7.1, it can be inferred that a quantification of positional errors will require knowledge of true positional values. Redundant measurements may be able to pinpoint the probable intervals for errors and thus the unknown truth, assuming some distribution for the positional errors. Alternatively, when expected positions for end points of a line segment are obtainable in some way, it will be possible to derive the expected position for an intermediate point x_r, using the linear combination suggested by Equation 7.1.

The variation of positional errors is usually measured by variance and covariance. This is done by applying the law of variance and covariance propagation, assuming the variation at the end points is known, and this is a classical subject in surveying adjustment and position fixing (Drummond, 1995; Shi, 1998; Burkholder, 1999). Based on the linear combination of random variables suggested in Equation 7.1, it is possible to derive a covariance matrix $COV(\varepsilon(x_r))$, of dimensions 2 by 2 for point x_r as:

$$COV(\varepsilon(x_r)) =$$

$$\begin{vmatrix} 1-r & 0 & r & 0 \\ 0 & 1-r & 0 & r \end{vmatrix} \begin{vmatrix} COV(\varepsilon(x_1)) & COV(\varepsilon(x_1),\varepsilon(x_2)) \\ COV(\varepsilon(x_1),\varepsilon(x_2)) & COV(\varepsilon(x_2)) \end{vmatrix} \begin{vmatrix} 1-r & 0 \\ 0 & 1-r \\ r & 0 \\ 0 & r \end{vmatrix} \tag{7.2}$$

which can be expressed in terms of the variances of the X and Y coordinate components of error vector $\varepsilon(x_r)$ and the covariance between X and Y as the matrix:

$$\begin{vmatrix} (1-r)^2\sigma_{X1}^2 + 2r(1-r)\sigma_{X1X2} + r^2\sigma_{X2}^2 & (1-r)^2\sigma_{X1Y1} + r(1-r)(\sigma_{X1Y2} + \sigma_{X2Y1}) + r^2\sigma_{X2Y2} \\ (1-r)^2\sigma_{X1Y1} + r(1-r)(\sigma_{X1Y2} + \sigma_{X2Y1}) + r^2\sigma_{X2Y2} & (1-r)^2\sigma_{Y1}^2 + 2r(1-r)\sigma_{Y1Y2} + r^2\sigma_{Y2}^2 \end{vmatrix}$$

where $\sigma_{X1}^2, \sigma_{Y1}^2, \sigma_{X2}^2, \sigma_{Y2}^2$, σ_{X1Y1}, σ_{X1X2}, σ_{X1Y2}, σ_{X2Y1}, σ_{Y1Y2} and σ_{X2Y2} stand for the variances and covariances of the X and Y components of the positional errors at the points as identified by the subscripts.

Suppose that the positional errors at the vertices have uniform variances σ_X^2, σ_Y^2 and covariance σ_{XY}. Further assume that positional errors are not correlated among neighbouring points. Then it is possible to derive a simplified version of the covariance matrix for the positional vector at point x_r as:

$$COV(\varepsilon(x_r)) = \left[(1-r)^2 + r^2\right] \begin{vmatrix} \sigma_X^2 & \sigma_{XY} \\ \sigma_{XY} & \sigma_Y^2 \end{vmatrix} \tag{7.3}$$

If normality in positional errors is assumed, as Leung and Yan (1999) were willing to do, it is possible to derive a probability density function based on the error quantities and covariances shown in Equations 7.1 and 7.3 respectively. Since a linear combination of normal variables is also a normal variable, it is straightforward to write the normally distributed probability density function for positional error at point x_r as:

$$pdf(x, x_r) = \left[2\pi\left((1-r)^2 + r^2\right)\left(\sigma_X^2\sigma_Y^2 - \sigma_{XY}^2\right)^{0.5}\right]^{-1} \tag{7.4}$$

$$\exp\left[-\left(\varepsilon_X(x_r)^2\sigma_Y^2 - 2\varepsilon_X(x_r)\varepsilon_Y(x_r)\sigma_{XY} + \varepsilon_Y(x_r)^2\sigma_X^2\right)\Big/ 2\left((1-r)^2 + r^2\right)\left(\sigma_X^2\sigma_Y^2 - \sigma_{XY}^2\right)\right]$$

where $\varepsilon_X(x_r)$ and $\varepsilon_Y(x_r)$ represent projected components of the error vector $\varepsilon(x_r)$ at point x_r along the X and Y coordinate directions, which are valued as the difference between a point $x(X,Y)$ and point x_r with coordinates (X_r,Y_r), that is:

$$\left(\varepsilon_X(x_r)\varepsilon_Y(x_r)\right)^T = x - x_r = (X - X_r, Y - Y_r)^T \tag{7.5}$$

where position $(X_r,Y_r)^T$ for x_r is given by evaluation against Equation 7.1.

The pdf for positional error at point x_r along the line segment x_1–x_2 is specified as a function of the location of point x with respect to x_r, as suggested by Equation 7.4. Such a specification enables an integral over x_r, reflecting the infinity of points along the line segment x_1–x_2. This integral, when followed by a division by the length of the line segment x_1–x_2, leads to a so-called line probability density function dependent on position x:

$$pdf(x) = \frac{1}{\|x_1 - x_2\|} \int_{x_1}^{x_2} pdf(x, x_r) dx_r \tag{7.6}$$

where the symbol $\|.\|$ stands for the Euclidean norm of distance. The denominator of line segment length serves to ensure that the integral of pdf(x) over the space is one, and a simple multiple integral rule can be used to verify this. Extensions to polylines and curves are possible, in principle, by integrating pdf(x,x_r) along the lengths of lines or curves.

It is interesting to illustrate some examples for point- and line-based probability density functions for positional errors in points and line segments. Error ellipses and epsilon error bands are widely known in GIS communities as useful descriptors for positional errors in point and line objects. Contoured versions of hypothetical probability density functions for errors in a point x_r and a line segment x_1–x_2 are shown in Figure 7.7. In Figure 7.7(a) and (b), dashed isolines correspond to different confidence levels in assessing the closeness between the measured and the true positions of point x_r and line segment x_1–x_2, respectively.

The linear combination suggested by Equation 2.34 has been applied flexibly for deducing analytical expressions for positional errors of lines. A further use can be made in deriving mean lines from a set of randomly distributed lines. The mean is an important statistical concept, and is meant to measure the average of a certain set of scalar data. Although the calculation of means is mathematically simple and easily programmable, their extension into the domain of discrete objects is far from straightforward (Goodchild *et al.*, 1995).

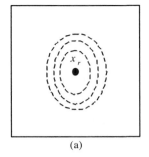   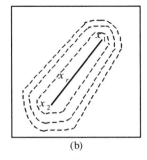

(a)   (b)

Figure 7.7 Error descriptors: (a) error ellipse; and (b) epsilon error band.

Consider the case of a line segment connecting two end points, whose generalisation to a union of line segments is possible. Suppose two points x_1 and x_2 are expressed as vectors $(X_1, Y_1)^T$ and $(X_2, Y_2)^T$, where the Xs and Ys stand for X and Y coordinates, respectively. The mean point x_m may be written as a vector of mean coordinates $(X_m, Y_m)^T$, that is, $X_m=(X_1+X_2)/2$ and $Y_m=(Y_1+Y_2)$, or more concisely as:

$$x_m = (x_1 + x_2)/2 \tag{7.7}$$

Consider two line segments L_1 and L_2, which are defined with end points x_{11}, x_{12} and x_{21}, x_{22} respectively. They may be represented as linear combinations $(1-r_1)x_{11}+r_1x_{12}$ and $(1-r_2)x_{21}+r_2x_{22}$, where scalars r_1 and r_2 are real numbers ranging between 0 and 1. Let x_{m1} and x_{m2} stand for the means of the end points corresponding to line segments L_1 and L_2, calculated from Equation 7.7. Then, the mean line segment $x_{m1}-x_{m2}$ may be expressed as $(1-r_m)x_{m1}+r_m x_{m2}$, where scalar r_m is again a real-numbered scalar in the interval between 0 and 1.

Extension to the case of more than two line segments is straightforward by averaging the respective number of end points. An example is provided in Figure 7.8, where the mean line and different realisations of lines are drawn as solid and dashed lines, respectively, with clusters of dispersed points circled.

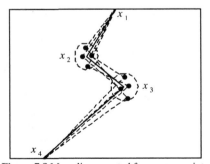

Figure 7.8 Mean lines created from mean points.

However, recovering mean lines in terms of Equation 7.7 depends on the correct definition of the starts and ends of the two line segments, which are, unfortunately, arbitrary, especially when complex line structures are concerned. This is a potential problem resulting from the use of discrete object-based vector data, preventing a generalised use of the mean line concept.

Of greater concern is the issue of spatial dependence. The preceding discussion has implicitly focused on modelling error in a single line segment. Even with this simplification, and with the assumption of spatially uncorrelated errors in the points comprising the lines, the numerical calculation of line density outlined by Equations 7.5 and 7.6 appears far from convenient. But realistic situations often involve non-zero spatial correlation in positional errors among neighbouring points and lines. Thus, the task of modelling positional errors and their propagation becomes overwhelmingly complicated when trying to accommodate spatial dependence within complete object representations.

The derivation of probability density functions for positional errors does not give a complete answer to the question of modelling errors in length and area estimates derived from points, even given knowledge of positional errors. This question relies, again, on the quantification and incorporation of spatial correlation in positional errors for estimating variances of length and area quantities, as will be described below.

Consider polyline objects that are sequences of line segments connecting vertices. Then, it is possible to sum the lengths of individual line segments to obtain the length of a polyline. If a polyline links three vertices (points) x_1, x_2, and x_3, it is straightforward to calculate its length as the sum of Euclidean distances $\|x_2-x_1\|$ and $\|x_3-x_2\|$.

The focus here is on deriving a variance estimate for polyline length. Refer to the formula for line segment length $\|x_{i+1}-x_i\| = [(X_{i+1}-X_i)^2+(Y_{i+1}-Y_i)^2]^{1/2}$, in which (X_i,Y_i) and (X_{i+1},Y_{i+1}) are the X and Y coordinates for points x_i and x_{i+1}, respectively. One can differentiate polyline length, $\|x_2-x_1\|+\|x_3-x_2\|$, with respect to X_1, Y_1, X_2, Y_2, X_3, and Y_3 as:

$$d(length) = \|x_2 - x_1\|^{-1}\left(-\Delta X_1 dX_1 - \Delta Y_1 dY_1\right) + \left(\|x_2 - x_1\|^{-1}\Delta X_1 - \|x_3 - x_2\|^{-1}\Delta X_2\right)dX_2$$
$$+ \left(\|x_2 - x_1\|^{-1}\Delta Y_1 - \|x_3 - x_2\|^{-1}\Delta Y_2\right)dY_2 + \|x_3 - x_2\|^{-1}\left(\Delta X_2 dX_3 + \Delta Y_2 dY_3\right) \tag{7.8}$$

where ΔX_i and ΔY_i stand for the shifts of X and Y coordinates from point x_i to point x_{i+1}, that is, $(\Delta X_i,\Delta Y_i)^T = x_{i+1}-x_i$.

Assume non-zero covariances between X and Y coordinates at individual points, and between X or Y coordinates in immediately neighbouring points such as x_1 and x_2, or x_2 and x_3. Then, on the basis of Equation 7.8, it is possible to derive the variance of polyline length from:

$$\text{var}(length) = \|x_2 - x_1\|^{-2}\left[\begin{array}{l}\Delta X_1^2\left(\sigma_{X1}^2 - 2\sigma_{X1X2} + \sigma_{X2}^2\right) + 2\Delta X_1\Delta Y_1\left(\sigma_{X1Y1} + \sigma_{X2Y2}\right) \\ + \Delta Y_1^2\left(\sigma_{Y1}^2 - 2\sigma_{Y1Y2} + \sigma_{Y2}^2\right)\end{array}\right]$$

$$+ \|x_3 - x_2\|^{-2}\left[\begin{array}{l}\Delta X_2^2\left(\sigma_{X2}^2 - 2\sigma_{X2X3} + \sigma_{X3}^2\right) + 2\Delta X_2\Delta Y_2\left(\sigma_{X2Y2} + \sigma_{X3Y3}\right) \\ + \Delta Y_2^2\left(\sigma_{Y2}^2 - 2\sigma_{Y2Y3} + \sigma_{Y3}^2\right)\end{array}\right]$$

$$- 2\|x_2 - x_1\|^{-1}\|x_3 - x_2\|^{-1}\left[\begin{array}{l}\Delta X_1\Delta X_2\left(\sigma_{X2}^2 - \sigma_{X1X2} - \sigma_{X2X3}\right) \\ + \Delta Y_1\Delta Y_2\left(\sigma_{Y2}^2 - \sigma_{Y1Y2} - \sigma_{Y2Y3}\right) \\ + \left(\Delta X_1\Delta Y_2 + \Delta X_2\Delta Y_1\right)\sigma_{X2Y2}\end{array}\right] \tag{7.9}$$

where $\sigma_{X1}^2, \sigma_{Y1}^2, \sigma_{X2}^2, \sigma_{Y2}^2, \sigma_{X3}^2, \sigma_{Y3}^2, \sigma_{X1Y1}, \sigma_{X1X2}, \sigma_{Y1Y2}, \sigma_{X2Y2}, \sigma_{X2X3}, \sigma_{Y2Y3}$, and σ_{X3Y3} stand for the variances and covariances of the X and Y components of the positional errors at or between the points as identified by the subscripts.

Equation 7.9 includes many covariance components for a simple polyline consisting of three vertices. It may be further expanded into a full list of covariances among a larger number (more than three as above) of neighbouring vertices comprising the whole polyline object under study. This implies rapidly growing difficulties in obtaining rigorous analytical expression of variance propagation from spatially varied and correlated positional errors as the number of points increases.

But if one is willing to assume zero spatial correlation in positional errors among neighbouring points, plus uniformity in X and Y variances and zero XY covariance, then simplification is possible. Thus, for a single line segment, Equation 7.9 is reduced to a simpler expression for variance in length $\|x_1-x_2\|$ as twice the variance in positional error. For a polyline that links up more than one line segment, variance in total length will depend not only on the number of line segments but also on whether these line segments are joined with little or striking curvatures. For a polyline of two line segments, the variance in length will be $(4 - 2\|x_2-x_1\|^{-1}\|x_3-x_2\|^{-1}(\Delta X_1\Delta X_2+\Delta Y_1\Delta Y_2))$ times variance in positional errors, as can be seen from Equation 7.9. A polyline that is smoothly curved will lead to a maximum value of $\Delta X_1\Delta X_2+\Delta Y_1\Delta Y_2$ but minimum curvature, and hence to minimum variance in line length. A polyline linking line segments that intersect one another at right angles will have a variance twice that of positional error multiplied by the number of line segments. For a polyline that consists of line segments that intersect one another with sharp angles, length will have a variance that increases steeply with the number of vertices, approximating $4n-6$ times positional error variance, with n being the total number of vertices ($n \geq 2$).

Consider now the case of area measurements of polygons. A combination of differentiation and variance propagation leads to the evaluation of variance in area estimation. Consider a polygon consisting of a sequence of vertices that are traversed counter-clockwise. Then, the areal extent of the polygon is calculated as a positive sum of vector products, as discussed in Chapter 2. The differential of the area of a polygon consisting of n vertices is given by:

$$d(area) = \frac{1}{2}\sum_{i=1}^{n}[(Y_{i+1} - Y_{i-1})dX_i - (X_{i+1} - X_{i-1})dY_i] \qquad (7.10)$$

where (X_i, Y_i) stand for the X and Y coordinates of the ith vertex numbered in a circular fashion (Chrisman and Yandell, 1988).

Now that the differentials of area have been obtained, area variance may be easily computed by propagation of the variances in positional errors of the component vertices of the polygon. As implied in the derivation of Equation 7.9, a complete listing of all variances and covariances between component vertices would be tedious. However, it seems reasonable to assume only local covariance within the three neighbouring vertices implied in Equation 7.10. This enables a simpler calculation of variance in polygon area:

$$\text{var}(area) = \frac{1}{4}\sum_{i=1}^{n}\left(\begin{array}{l}\Delta Y_{i+1}^2\sigma_{X_i}^2 + \Delta X_{i+1}^2\sigma_{Y_i}^2 - 2\Delta X_{i+1}\Delta Y_{i+1}\sigma_{X_iY_i} \\ + 2\Delta Y_i\Delta Y_{i+1}\sigma_{X_iX_{i+1}} + 2\Delta X_i\Delta X_{i+1}\sigma_{Y_iY_{i+1}}\end{array}\right) \qquad (7.11)$$

where $\sigma_{X_i}^2$, $\sigma_{Y_i}^2$, $\sigma_{X_iY_i}$, $\sigma_{X_iX_{i+1}}$, and $\sigma_{Y_iY_{i+1}}$ stand for the variances and covariances of the X and Y components of the positional errors at the vertices, as indicated by the subscripts, and column vectors $(\Delta X_i, \Delta Y_i)^{\mathrm{T}} = x_{i+1} - x_{i-1}$, and $(\Delta X_{i+1}, \Delta Y_{i+1})^{\mathrm{T}} = x_{i+2} - x_i$, representing the shifts of X and Y coordinates between specified points.

Further simplification may be made by assuming zero spatial correlation between neighbouring vertices defining the polygon's boundaries, plus uniform variances in X and Y coordinates but zero covariance. This leads to an evaluation of standard error in polygon area as half the standard error in points, multiplied by the square root of the sum of squared distances of the component line segments defining the boundaries of the polygon.

This implies that the area calculated for a polygon defined with discrete vertices could have worryingly high levels of uncertainty, since the standard error in the calculated area would become unacceptable, in particular, with elongated polygons with large ratios of perimeters to areal extents. On the other hand, variances in area estimates are reduced for polygons with compact boundaries.

The accuracy of length and area calculation, both of which are classic GIS functions, is normally assumed to be acceptable. Analytical approaches to modelling positional errors and their impacts on common measurements for line and area objects are far too complicated for general users of vector GISs. But simplification of error propagation by assuming spatial independence tends to suggest unrealistic uncertainties in derivative data products. Alternative strategies of greater flexibility would be particularly welcome.

By applying stochastic simulation, one can deal with the spatial dependence that is intrinsic to fields of positional errors. And with simulated fields of positional errors, it is easy to apply statistical techniques for deriving error parameters such as means and standard deviations for a geo-processed data set. In this way, one obtains robust operational definitions of mean lines and epsilon error bands from simulated fields of positional errors, and a sound basis for analytical approaches. Moreover, simulation will permit flexible quantification of variances in derivative data sets such as line length and polygon areas, as discussed in the next section.

7.3.2 Stochastic Simulation of Positional Errors

Discrete objects are useful representations for simple points such as the ground-measured GCPs used for orienting a photographic pair, and for real entities, such as roads and buildings, that are digitised from large-scale plans. For discrete objects that are unambiguously defined, emphasis is placed on positional errors. Vonderohe *et al.* (1990) performed positional accuracy assessment for a set of digital cadastral data by using statistical comparisons. But further work is clearly needed to provide information on uncertainty in complex objects.

Section 7.2 proposed a general framework for describing object-related uncertainties, and the previous section discussed the analytical modelling of positional errors and their propagation into derivative measurements such as length and area. As suggested, the task of error modelling in objects composed of vertices becomes very complex if an analytical approach is taken. Thus, it may be better to pursue stochastic simulation.

Error ellipses are suitable models as well as descriptors of positional errors in points. This is because, given parameters for an error ellipse, one is able to generate distorted point samples, which conform to the distribution prescribed by the given parameters. For lines, the situation becomes more complicated. Epsilon error bands, though useful as good error descriptors, do not satisfy the requirements of error models on their own. The reason is that, unlike an error ellipse model that is fully defined with variances and covariances in and between X and Y coordinates, the provision of an error band width alone does not provide any information on the significant spatial correlations between errors in the points comprising the line. Besides, relationships between the geometry of a line's error band and defining error ellipses in its component points are far from simple, and band width is often an over-generalisation of uncertain lines. Therefore, the quantification of error bands does not generally allow for simulating equal-probable lines, unless errors in points comprising lines are assumed to be independent of their neighbouring points so that lines are simulated through simple distortion of points.

On the other hand, it is relatively easy to simulate equal-probable lines with properly incorporated and spatially correlated point errors by computer. Alternative lines are generated with vertices distorted on the basis of appropriately simulated positional errors. Further, epsilon error bands can be derived by overlaying simulated lines, and represented graphically with isolines of confidence.

Thus, lines should be more usefully defined as stochastic line realisations, which are topologically structured from simulated vertices with specific levels of spatial correlation in local errors. Multiple realisations of lines may occur in the manner shown in Figure 7.8, where line segments connecting different vertex realisations follow certain stochastic characteristics which have been deemed significant for the data set under study.

The displacements imposed on positional data by simulation must make sense in a specific problem domain. For example, Fraser and Shortis (1992) discussed an empirical correction model for radial distortion in photographic fields, which may shed light on constructing fields of positional errors based on experimentation. Tudor and Sugarbaker (1993) described a method for orthographic digitising of aerial photographs utilising DEM data. If positional errors are believed to arise due to errors in the DEM used for digital image rectification, then the simulation of the former may benefit from exploiting photogrammetric imaging equations and DEM errors.

Computer scientists have contributed to the discussion of GIS errors by introducing the methods of deformable templates and deformable contours (Grenander, 1996; Jain *et al.*, 1996). A deformable template is constructed as a two-dimensional displacement field of a continuous function with certain boundary conditions. Kiiveri (1997) was able to extend the use of a deformable template with a trigonometric basis into the simulation of distorted polygonal data. However, spatial correlation in positional errors is rarely incorporated in such deformable templates in an explicit way. Moreover, there are no easy rules to follow for quantifying the spatial correlation that exists in real data sets.

Autoregressive methods have found useful applications in the description of cartographic boundaries based on curvatures, making use of local correlation in curvature data (Garcia *et al.*, 1995). Such a method can provide grounds for using a simplified formula for area variance, as in Equation 7.11. As Haining *et al.* (1983) explained, spatially autoregressive fields are usually generated in an iterative way so that the resulting surface recovers the desired degree of spatial autocorrelation. Usually, pseudo-random numbers are generated by a computer (Hunter and Goodchild, 1997). Geostatistical approaches to simulation are attractive, as they are able to quantify and incorporate the degree of spatial correlation intrinsic to a particular data set, leading to a data-driven solution.

Simulation is better performed using raster fields and depicting errors as the variables. Applying raster structures for vector data error seems questionable at first, as vector and raster are usually treated in GIS as two distinctive data structures. The use of raster structures for vector data is justified for reasons illustrated by the following arguments. As shown in Figure 7.6, positional errors lead to non-sharp boundaries for areas. When dealing with pure points and lines, as opposed to areas, the results may appear unusual. In other words, the use of the logic of lines and points defined by real numbers leads to intriguing situations. It is not difficult to deduce that the probability of precisely locating a single point, say point x_r in Figure 7.6, vanishes to zero, since including or excluding a single point makes no difference to the probability of locating a position x within a certain locality. The same result applies to lines of zero width (i.e., geometric lines), as the convolution operation leading to Figure 7.6(c) is, in fact, a two-dimensional calculation. However, one might have anticipated that the probability of placing points exactly at a location or along a specified line should be positive, if sufficient accuracy is maintained in measuring a point or tracking a line.

To resolve these problems in evaluating the probability of locating pure points and lines, it is worth recalling that real-world points and lines have finite sizes. In other words, pure points and lines do not exist in reality and polygonal boundaries have to be simplified to cater for digital processing of spatial data. For example, map lines may represent roads and rivers of finite width, and shorelines of lakes on maps are commonly approximated by generalised polylines. Even well-defined point and line objects measured with acceptable accuracy will have effective sizes due to positional errors, leading to the existence of objects of finite sizes. It will be more objective and sensible to conceive of a vector data set as a raster of variable resolution associated with positional errors, than as discrete geometric primitives of spurious accuracy. Such a raster-based strategy will ensure that zero probability ceases to occur at raster cells on or containing individual objects, even within an extremely narrow window of a point, as shown in Figure 7.2. Therefore, the introduction of a raster structure whose equivalent resolution is adapted to the inherent positional error in the underlying vector-structured data, is well justified as a move to accommodate positional errors and to permit flexibility in geo-processing.

Further, simulation is more concerned with reproducing spatial variability than with getting the best local estimates for the variables under study. Thus, the approximation involved in the transformation between raster and vector is a trade-off for mapping spatial variability and dependence in positional errors, which are essential to objective assessment of the impacts of positional errors upon composite lines and areas. Dunn *et al.* (1990) provided empirical evidence that there is a certain parallel between raster data and vector data of limited accuracy.

To conduct a stochastic simulation, the spatial correlation in the underlying process or properties (attributes) must be quantified and modelled. In geostatistics, the properties of interest are modelled as spatially correlated random variables, which can take a variety of outcome values according to some probability distribution. Positional errors $\varepsilon(x)$ can be viewed as spatially varied but correlated vectors. Under the intrinsic hypothesis, a stationary variance $\gamma(h)$ of the differences in error vectors between places separated by a given distance and direction is assumed. Funk *et al.* (1999) applied semivariogram modelling to the angular measurements of positional error vectors in the context of urban road networking, assuming a global measure for positional error magnitude.

Consider the distribution of a positional error vector $\varepsilon(x)$ over a field A, $\{\varepsilon(x), x \in A\}$. Stochastic simulation is the process of building alternative, equally probable, high-resolution models of $\varepsilon(x)$, with each realisation denoted with the superscript τ: $\{\varepsilon^\tau(x), x \in A\}$. Conditional simulation results in realisations that honour the data values of observations $\varepsilon(x_s)$, that is, $\varepsilon^\tau(x_s) = \varepsilon(x_s)$, where subscript s stands for the label of a point with measured positional error. Positional errors are two-dimensional vectors, as shown earlier in this chapter. However, it may be reasonable for positional errors to be simulated in X and Y coordinates independently. These simulated X and Y components are then joined to recover simulated error vectors for individual locations.

Figure 7.9 illustrates a hypothetical example where a line segment shown in Figure 7.9(a) is distorted through displacing its end points following a Gaussian disturbance of 1 m variance. A raster cell size of 0.1 m was adopted and Bresenham's line drawing algorithm was used to tessellate simulated line segments over the raster plane (Bresenham, 1965). Positional errors were simulated 1000 times for the two end points before the distorted versions of line segments were rasterised and summed to produce the probability density surfaces of the line segment shown in Figure 7.9(b) and (c).

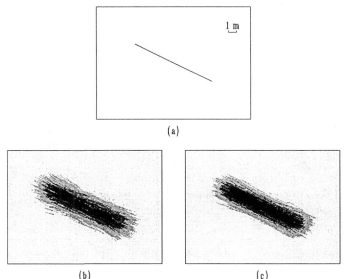

(a)

(b)   (c)

Figure 7.9 A line segment (a) and surfaces of probability density
simulated with random (b) and spatially correlated (c) positional errors.

In Figure 7.9(b), end points are taken independently, resulting in shaded error bands typically bent inwards at the middle but much dispersed at the two ends. On the other hand, as shown in Figure 7.9(c), when errors at the end points are spatially correlated with a Gaussian model of a unit sill and a range three times the line segment length, shaded error bands appear tighter than those shown in Figure 7.9(a).

Figure 7.9 provides evidence for a flexible solution to quantifying line probability densities using stochastic simulation, since positional errors in component vertices are often spatially correlated rather than purely random. Such a stochastic numerical solution serves also to remove the confusion that has cloaked the use of epsilon bands for the description of line errors. Without the simulation protocol, Figure 7.2(b) would still be a mere graphic of error zones around a line segment, and would be hard to verify analytically using Equation 7.6.

Means and standard deviations can be easily derived on the basis of simulated positional errors, which can lead to straightforward quantification of line errors. Denote standard deviations in X and Y error fields, specific to location x, by $stdx(x)$ and $stdy(x)$. Then, a standardised measure of planimetric displacement localised at x may be calculated as the square root of the summed squares of $stdx(x)$ and $stdy(x)$:

$$\text{standard displacement}(x) = \sqrt{stdx(x)^2 + stdy(x)^2} \qquad (7.12)$$

This type of standardised displacement may be evaluated and averaged at locations of specific lines to generate estimates for the widths of epsilon error bands. Thus, simulation modelling provides a much easier way of mapping epsilon error bands.

Technical details of both conditional and unconditional simulation have been addressed in earlier chapters, and will not be discussed here further. But there are certain points peculiar to the simulation of positional errors. Usually, position and attributes are separable for well-defined spatial entities, implying that attributes can be handled independently of position. However, for positional errors, as special attributes of entities, simulation involves direct displacement of positions in points and lines, and there is no guarantee that topological structures are maintained in simulated data. Thus, torn or folded spaces must be avoided in distorted versions of original vector data (Goodchild, 1999a). In other words, there is a need in stochastic simulation to check for any topological inconsistencies that might occur in simulated positional errors in vector data.

Assume that topology is built appropriately for a vector data set before simulation. Topological consistency may be violated in simulated distortions of lines and polygons if folding occurs, leading to inappropriate intersections between distorted features. Figure 7.10 illustrates a situation where a fold results from simulated displacements $\varepsilon(x_1)$ and $\varepsilon(x_2)$ corresponding to two locations (i.e., grid nodes) x_1 and x_2, as indicated in the diagram.

While intersection of vectors can be detected from the method discussed in Chapter 2, a simpler implementation is possible with a raster structure. The displacements in Figure 7.10(b) can easily be checked for the occurrence of folds when projected to the X coordinate direction. The condition under which a fold is formed is:

$$\varepsilon_x(x_2) - \varepsilon_x(x_1) \le -\|x_2 - x_1\| \quad \text{or} \quad d\varepsilon(x)/\|x_2 - x_1\| \le -1 \qquad (7.13)$$

where $\|x_2-x_1\|$ is the size of a grid cell in the X direction, while $d\varepsilon(x)$ is the increment of values from node x_1 to node x_2 (i.e., $d\varepsilon(x)=\varepsilon(x_2)-\varepsilon(x_1)$, with positive values of $\varepsilon(x_1)$ and $\varepsilon(x_2)$ along the X direction being shown in Figure 7.10(b).

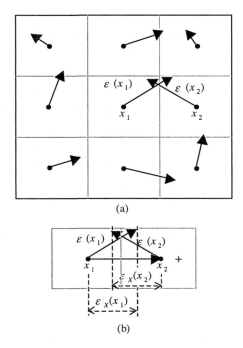

(a)

(b)

Figure 7.10 Occurrence of folds: (a) displacement vectors; and (b) the condition when a fold forms.

The condition indicated in the inequality (Equation 7.13) confirms that a fold is formed if the difference of simulated values of two neighbouring grid cells is no greater than the negative of the grid cell size along that direction. In other words, folds tend to happen when the slope of the change in neighbouring positional errors is too great in the direction opposite to that of the positional error. Shown in Figure 7.11 is an example in which a line and a polygon close to each other intersect after simulation, a typical result of folding. One might imagine the line and polygon shown in Figure 7.11 as representing a section of coastline and an island, respectively. The intersection illustrated in Figure 7.11(b) would be an apparently erroneous representation of reality.

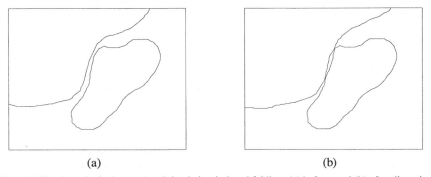

(a)   (b)

Figure 7.11 A hypothetical example of simulation-induced folding: (a) before; and (b) after distortion.

Folding may be avoided by constraining simulation. One rule is to work with positional errors that vary continuously and smoothly over space with strong spatial correlation (Goodchild, 1999a). Thus, as opposed to the condition set in Equation 7.13, this will allow the increment of positional errors, $\varepsilon(x+dx)-\varepsilon(x)$, to tend to 0 as the positional increment (or lag) dx approaches 0. This is not unreasonable, since many variables are assumed to be continuous and correlated in order to permit locations to be interpolated with reasonable accuracy, and to allow the use of rubber-sheeting for map or image registration. A related consideration is to discretise fields of positional errors into raster systems of resolutions compatible with the underlying positional errors. This will prevent folds from occurring simply because raster cell sizes are set too fine in relation to the magnitude of positional errors observed.

There is yet another concern for simulating positional errors, which is about identifying point pairs corresponding to before- and after-distortion, which allow for mapping fields of displacement vectors, if simulation conditional to a particular data set is required. Filin and Doytsher (1999) proposed a linear mapping approach to matching polylines for purposes of map conflation, generalisation, and change detection, which may be usefully adapted for matching vertices and nodes across map layers. Comprehensive research was carried out by Noronha *et al.* (1999) and Noronha (2000) in the context of intelligent transportation navigation. When many different layers of positional data with disparate levels of accuracy are overlaid, there tend to be significant ambiguities with respect to matching points or lines, even when they are actually identified correctly on individual map layers. If objects are measured with different accuracy, or located in different reference systems, identical objects may be missed in match detection, or different objects may be mistakenly equated.

A useful method for robust matching of objects is to set up a hierarchy of accuracy standards for different classes of objects. Thus, a road in a rural area can be captured from the field using GPS with a lower accuracy standard, and yet this will pose no problems for its identification in GIS, whereas road networks in densely urbanised areas have to be measured with much higher accuracy in order not to be confused. Further enhancement is possible by exploring attribute data in object-based GIS, so that semantic information may be utilised in object identification.

Despite such considerations regarding implementation, there are no major difficulties in conducting stochastic simulation on spatial databases. With appropriate use of such methods, it is generally possible for geographical information scientists and decision-makers to place confidence limits on a particular piece of derivative information inferred from data sources by geo-processing.

7.4 EXPERIMENTING WITH ERROR MODELLING IN VECTOR DATA

7.4.1 Test Data

As described in earlier chapters, there is a need in suburban applications to combine data sets with quite different positional and attribute accuracies. In addition to elevation and land-cover data, the mapping of suburban areas involves a range of medium- and fine-scale data products, which usually incorporate precise positional data for such features as buildings, street networks, and land records. This section complements error handling for continuous and categorical data with a realistic test of errors in positional data.

At the Edinburgh suburb test site, the 1:24,000 scale aerial photographs were used to generate test data, and 1:5,000 scale aerial photographs to provide reference data. Ground control for the aerial photographs was established by land surveying, which, in turn, enabled the implementation of a photogrammetric block adjustment using the 1:5,000 scale aerial photographs. This block consists of two strips of photographs, each of five stereo-models, generating photocontrol for photogrammetric digitising based on the 1:5,000 scale aerial photographs. One stereo-model using the 1:24,000 scale aerial photographs was oriented using ground control points obtained jointly by the land surveying and the block adjustment (Zhang, 1996).

Photogrammetric digitising was then performed using an AP190 analytic plotter. The spatial entities extracted include property boundaries formed by hedges and walls, houses and buildings, footpaths, and the margin of a large pond. Both the test data and reference data were transformed into ARC/INFO coverages for ease of management and data analysis. The vector data are shown as maps in Figure 7.12, where thin and bold lines are based on the 1:24,000 scale and 1:5,000 scale aerial photographs, respectively.

To provide initial estimation of the positional accuracy of points digitised from the 1:24,000 scale aerial photographs, measured coordinates of a few well-defined points were checked against the points densified from the block adjustment based on the 1:5,000 scale aerial photographs. Common examples were building corners, centres of manholes, and road junctions. The validation points were not those used in controlling the 1:24,000 scale stereo-model, so that the condition of independence of validation data is not violated. Results are listed in Table 7.1, with RMSEs of the planimetric (X,Y) data (ASPRS, 1989).

As indicated in Table 7.1, positioning by photogrammetric digitising is accurate at a RMSE of 0.6 m in both X and Y coordinates, which may be sufficient for many purposes such as urban management, traffic planning, and certain engineering applications. However, this level of accuracy must be interpreted as being relevant to individual, well-defined points only, and extrapolating it to more complex objects is subject to greater uncertainty. For instance, it would be meaningless to evaluate, in terms of the RMSE measures reported in Table 7.1, the positional accuracy of objects with indeterminate footprints.

Even for discrete lines and polygons, the use of point-based RMSEs would be limited, because geographical lines and polygons can rarely be interpreted as consisting of independently positioned points on a simple geometric basis, as explained by Zhang and Kirby (2000). In other words, spatial dependence is often significant between positional errors in neighbouring points comprising lines and polygons, suggesting that accuracy evaluation for lines and polygons is far from straightforward.

Stochastic simulation works better than analytical approaches by simulating alternative versions of a vector data set so that it becomes easier to predict the accuracy of derivative data products, such as line lengths and polygon areas, as discussed in Section 7.3. In the stochastic simulation of vector data, equally probable versions of the underlying data are generated that include positional errors consistent with what is known about the uncertainties in the data. With such simulated vector data, GIS users will be able to assess the consequences of using the data in certain map operations.

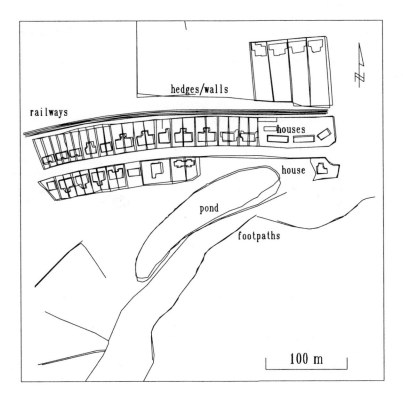

Figure 7.12 Photogrammetrically digitised vector data using the 1:24,000 scale aerial photographs (thin lines) and the 1:5,000 scale aerial photographs (bold lines).

Table 7.1 Accuracy estimate for photogrammetrically digitised points (unit: metres).

	Test	RMSE	
Model scale	Number of points	X	Y
1:24,000	15	0.595	0.569

7.4.2 Stochastic Simulation of Vector Data

In stochastic simulation, it is necessary to compute experimental semivariograms and then to create suitable semivariogram models. For this, the test vector data digitised from the 1:24,000 scale aerial photographs were overlaid with the reference data acquired from the 1:5,000 scale aerial photographs. Then, it was possible to match points (including nodes and line vertices) across the two data layers, and to compute the displacements for individual points. This process resulted in an extended ARC/INFO point attribute data (.PAT) file comprising IDs, reference coordinates, and the displacements for 785 correctly matched points.

Data were reorganised from the .PAT file to build a new data file suitable for GSLIB to calculate semivariograms. Experimental semivariograms for the positional errors in X and Y coordinates are shown in (a) and (b) of Figure 7.13, respectively.

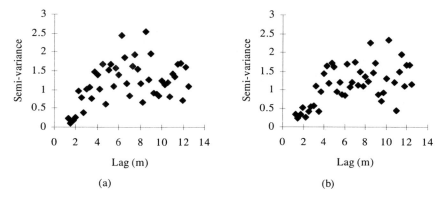

Figure 7.13 Experimental semivariograms for positional errors in coordinates: (a) X; and (b) Y.

In Gaussian model-based simulation, original error data must be transformed into normal scores. Experimental semivariograms were calculated for normal scores of positional errors in the X and Y coordinates. Results are shown as diamonds in Figure 7.14, where (a) and (b) correspond to the normal scores transformed from X and Y coordinate components of positional errors, respectively.

Fitted models are shown as lines in Figure 7.14, where the solid and dashed lines correspond to the Gaussian and spherical models respectively. With the Gaussian models, estimated values of the nugget effect, sill, and range were 0.15, 0.85, and 4.5 m for the X coordinates, and 0.2, 0.8, and 5.5 m for the Y coordinates, respectively.

A well-respected geostatistical rule is that semivariogram modelling is more crucial for capturing spatial dependence in any regionalised variables at smaller lags than at larger lags. Thus, the Gaussian and spherical models shown in Figure 7.14 should not be evaluated based on goodness of geometrical fit alone. Figure 7.14 shows that the Gaussian models display smoother slopes at small lags with non-zero nuggets, while the spherical models show a sharper increase in semi-variance at small lags. It may therefore be sensible to adopt the Gaussian model not only on the basis of the F statistics, but also because they will lead to smoother fields of positional errors. The characteristic of smoothness will help prevent erratic behaviours such as folding in simulated vector data, as described in the condition identified in Equation 7.13. Moreover, the incorporation of non-zero nugget effects means that micro-scale fluctuations in displacements will be treated as random errors, an approach that seems better suited to positional data that are known to possess measurement errors.

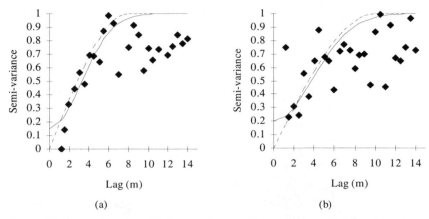

Figure 7.14 Semivariogram models for normal scores of positional errors in: (a) X; and (b) Y.

Stochastic simulation (both unconditional and conditional) was performed using a Gaussian sequential simulation program SGSIM provided in GSLIB. The parameter file was supplied with suitable data including grid cell size (8 m by 8 m), ranges, sills, and nugget effects describing the semivariogram models. For both modes of simulation, ten realisations were created for positional errors in X and Y independently, which were stored as new data items in the .PAT file mentioned above. From these simulated positional errors, ten versions of vector data were generated for both modes of simulation.

Shown in Figures 7.15 and 7.16 are realisations of distorted vector data, which were created from unconditional and conditional simulation of positional errors, respectively. Displayed in Figure 7.15(a), (b), and (c) are unconditional simulated error fields in X and Y, and the resulting coverage (in solid lines), respectively. Figure 7.15(c) indicates the discrepancy between the simulated and the original coverages digitised from the 1:24,000 scale aerial photographs (as dashed lines).

For comparison, a conditionally simulated coverage is shown in solid lines in Figure 7.16(c), where the error fields in X and Y are displayed in (a) and (b), respectively. Again, for the sake of comparison, the simulated coverage is overlaid with the original coverage (dashed lines) in Figure 7.16(c), where differences between both coverages are mapped.

As one might have anticipated, simulation in a conditional mode generates much smaller displacements than in an unconditional mode: the overlay shown in Figure 7.16(c) reveals much thinner slivers than that in Figure 7.15(c), and some line segments in the former appear to be coincident. This is because conditional simulation requires that simulated displacements at data points recover the observed (original) distortion values, leading to zero deviations from the original coverage at data points.

Topologic inconsistency was checked using the criterion expressed in

Equation 7.13. No folds occurred when the grid cell size was set at both 8.0 m and 4.0 m, because the distortion surfaces then generated were smooth enough in terms of the grid cell size. However, when the grid cell size was set at 2.0 m six folds were detected. Figure 7.15 shows a case of folding, where houses and property boundaries are too crowded to simulate distorted lines with the correct topology being maintained.

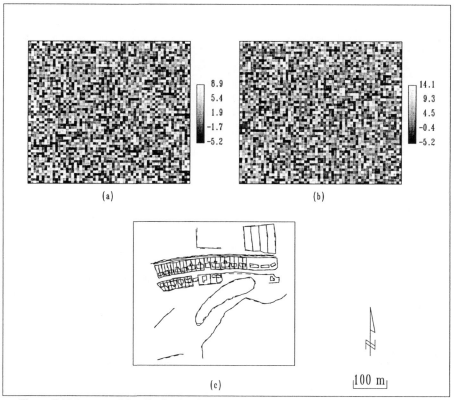

Figure 7.15 An unconditionally simulated coverage: (a) and (b) simulated error fields in X and Y respectively; and (c) the simulated data (solid lines) overlaid with the original data (dashed lines).

The fact that topological inconsistency (e.g., folding) occurred when the raster resolution was set at 2 m, the finest resolution tested, raises the issue of matching resolution (precision) to accuracy. The resolution of 2 m may be too fine to accommodate the spatial variation of positional errors in the original vector data, leading to an increased possibility of folding.

To avoid the occurrence of artefacts such as folds when performing computer simulation of error-prone vector data, knowledge of positional errors is important for setting resolutions that are coupled to the inherent levels of inaccuracy. Goodchild (1999a) discusses the issues of objects and positional errors. A feasible approach would be to adapt spatial resolution to positional inaccuracies. By doing so, finer resolutions would be used for more accurate data, and coarser ones for poorer data. It should be emphasised that the occurrence of topological inconsistency should not be seen as confined to raster-based simulation methods for vector data. Other simulation methods are not necessarily immune to this type of problem, as noted by Hunter and Goodchild (1996) and Kiiveri (1997).

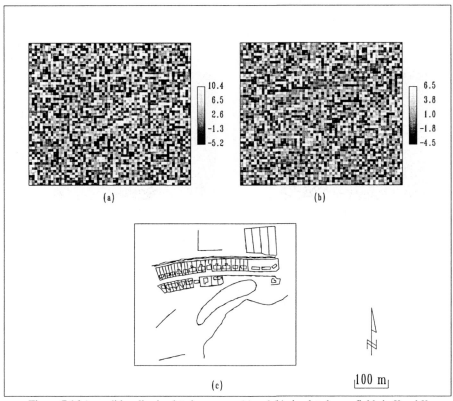

Figure 7.16 A conditionally simulated coverage: (a) and (b) simulated error fields in X and Y respectively; (c) the simulated data (solid lines) overlaid with the original data (dashed lines).

7.4.3 Deriving Mean Lines and Epsilon Error Bands from Simulated Data

Geostatistical simulation provides powerful tools for modelling uncertain spatial data, a claim that will be reinforced further in the remainder of this chapter. As discussed in Section 7.3, the modelling of errors in objects as well as fields using analytical approaches is prohibitively complex, a problem that discourages general GIS users from exploring error modelling due to daunting amounts of mathematical detail. It will be shown in the context of error modelling for vector data that mean lines (or areas), epsilon error bands, and errors in other object measurements are readily derived from simulated objects using methods that are conceptually simple, despite being at times computationally complex.

Once the simulation of positional errors is computerised, it is straightforward to derive fields of mean positional errors from simulated error fields. Results based on unconditional simulations are shown in Figure 7.17, where the mean positional errors in X and Y coordinates are displayed in (a) and (b) respectively, while (c) overlays the mean lines (solid lines) against the original data (dashed lines).

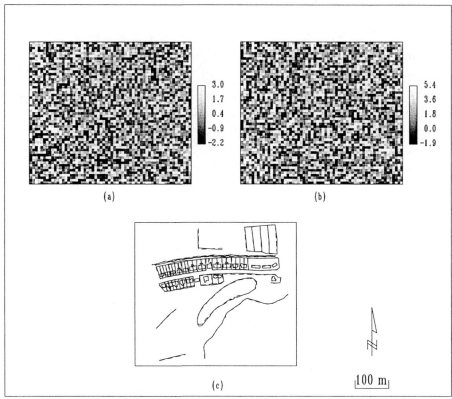

Figures 7.17 Mean lines derived from unconditionally simulated vector data:
(a), (b) mean positional errors in X and Y respectively;
(c) the mean lines (solid lines) overlaid with the original data (dashed lines).

Comparison of Figure 7.17 with Figure 7.15 reveals that there is obviously less spatial variability in mean error fields than in individually simulated error fields, as one might have expected. This is because the averaging effects induced by the mean operation suppress spatial variability in the simulated data, leading to much smoother fields of mean errors. Also, on the whole, the mean lines are shown to be closer to the original lines than the simulated lines, because mean lines represent spatial versions of mean quantities and should have generally reduced deviation from the assumed true values.

Similarly, mean errors fields and mean lines can be generated from the conditionally simulated data set, as shown in Figure 7.18. Mean error fields in X and Y are shown in (a) and (b), respectively, while the mean lines are displayed as solid lines in (c) in comparison with the original data (in dashed lines). Again, spatial variability in the mean error fields is reduced relative to that in individually simulated data, leading to smaller errors as reflected in the sizes of the slivers.

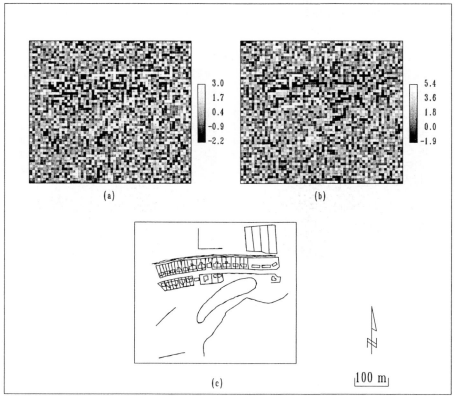

Figures 7.18 Mean lines derived from the conditionally simulated vector data:
(a), (b) mean positional errors in X and Y respectively;
(c) the mean lines (solid lines) overlaid with the original data (dashed lines).

It would be hard, if not impossible, to derive mean lines by analytical methods despite various simplifying assumptions, as has been discussed in Section 7.3. Stochastic simulation, on the other hand, forms a flexible approach using semivariogram modelling and appropriate spatial resolution to generate equally probable realisations of fields of positional errors. Simulated error fields can be used either to recover distorted versions of a vector data set or to quantify location-specific means and standard deviations efficiently, both of which tasks would be extremely complicated if approached directly in the domain of vector data using analytical methods.

In addition to the calculation of mean errors, standard deviations in fields of positional errors can also be derived from the simulated data. A measure of standardised planimetric displacement for individual locations can then be calculated as the square root of the summed squares of standard deviations in X

and Y coordinates, as formulated in Equation 7.12. This measure of planimetric displacement can be evaluated at line positions to derive estimates of epsilon errors. The set of error data was put into .AAT files as extended attributes for the polylines recorded in the original ARC/INFO coverage. Results for the unconditionally simulated vector data are shown in Figure 7.19. Standard deviations in X and Y positional errors are depicted in Figure 7.19(a) and (b), respectively, while the epsilon error bands are displayed in Figure 7.19(c) as buffered zones of width equal to twice the standard displacements.

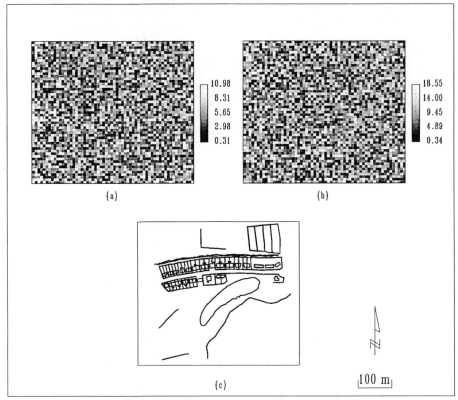

(a)

(b)

(c)

100 m

Figure 7.19 Epsilon errors derived from the unconditionally simulated vector data:
(a), (b) standard deviations in X and Y positional errors respectively; (c) epsilon error bands.

The set of computer-rendered maps in Figure 7.19 provides an effective visualisation of lines of finite width. Results for the conditionally simulated vector data are shown in Figure 7.20, where standard deviations in X and Y positional errors are depicted in (a) and (b), while the epsilon error bands are displayed in (c). Comparing Figures 7.19(c) and 7.20(c) indicates that unconditional simulation generates much larger error bands than conditional simulation. Not surprisingly, some polylines in the unconditionally simulated data are hardly visible, since the epsilon error bands that were used as parameters for buffering were not big enough in relation to the scale of the maps.

Enlarged displays of Figures 7.19(c) and 7.20(c) may be useful for visualising the spatially varying bands of line errors. Enlarged versions for Figures 7.19(c) and 7.20(c) are shown in Figures 7.21 and 7.22, respectively. These enlarged displays provide closer views of the error bands, which are drawn as filled buffer zones along the underlying lines.

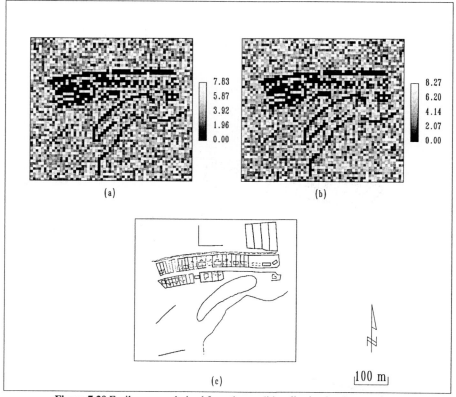

Figure 7.20 Epsilon errors derived from the conditionally simulated vector data:
(a), (b) standard deviations in X and Y positional errors respectively; (c) epsilon error bands.

As expected, conditional simulation of positional data implies that data locations should remain where they are not only on the original maps but also on the simulated maps. Since the majority of data points were used for conditional simulation, which required the fixing of data points, there should be little variation in the simulated data at data points.

However, the conditionally simulated data gave rise to an average standard planimetric displacement of about 0.8 m, which is significant. This level of inaccuracy was actually a result of artefacts of the raster structure used for the simulation of vector data. In terms of the positional accuracy shown in Table 7.1, which is also about 0.8 m, the inaccuracy imposed by raster artefacts should not be a major source of concern, but rather should be reckoned as comparable to the original data inaccuracy. Indeed, the existence of positional errors in vector data implies that one has to live with a digital world of finite resolution and accuracy. Thus, the raster-based methods for handling positional errors of vector origin should not be seen as an imposed limitation but as a sensible adaptation to reality. Such side effects of using raster structures for simulating vector data can be resolved by discounting them in the final assessment, so that only the net effects due to data errors are evaluated.

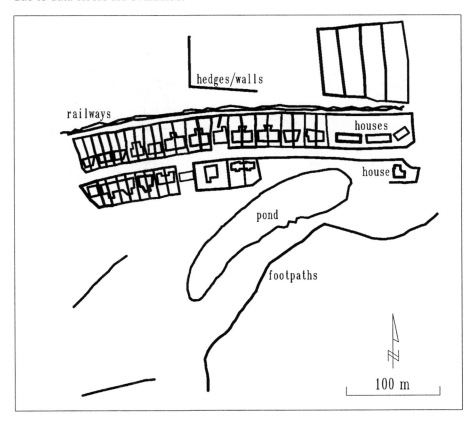

Figure 7.21 An enlarged display of epsilon errors derived from
the unconditionally simulated vector data shown in Figure 7.19(c).

In summary, the experiment described thus far focused on a geostatistical approach to modelling positional errors, and was built upon the capability and flexibility of raster structures to accommodate the spatial heterogeneity and correlation intrinsic to many spatially distributed phenomena. Raster structures were conceived of being composed of densely varying fields of positional errors, upon which vector data were overlaid and interpolated at data locations. In comparison, unconditional simulation seems more natural as a way to evaluate inaccuracies due to positional errors than conditional simulation. This is because conditional simulation often results in nothing more than the artefacts of rasterisation at data locations, if all component points in an object data set are used in the conditioning, unless random errors are added to generate distortions. Lastly, these approaches have explored an interesting common ground between raster and vector structures, where the two ways of organising data are collaborative rather than conflicting.

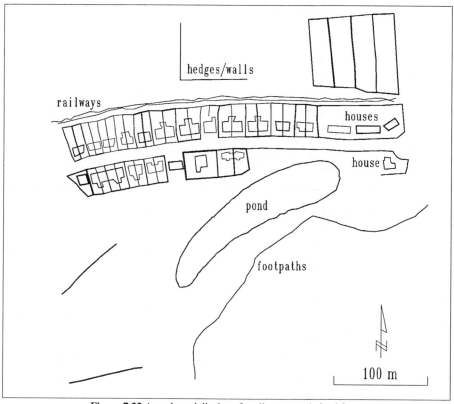

Figure 7.22 An enlarged display of epsilon errors derived from
the conditionally simulated vector data shown in Figure 7.20(c).

7.4.4 Error Statistics for Lines and Areas

Length and area calculation represents a classical yet important GIS function. Errors in length and areas calculated from vector data reflect propagated positional errors in the points and lines comprising the vector data set. How to predict errors in length and area estimates is a topic of continuing research (Chrisman and Yandell, 1988; Prisley *et al.*, 1989; Kiiveri, 1997).

In this experiment, attention was given to the estimation of perimeters and areas for the two polygons, the pond and a house, labelled in Figure 7.12. Variance propagation was based on the formulae in Equations 7.9 and 7.11. Assuming zero correlation among adjacent points and independence between X and Y coordinates, theoretical values of standard errors in perimeter and area estimates were computed for the two polygons based on the positional error estimates shown in Table 7.1.

In addition, ten versions of simulated vector data sets were examined with respect to the means and standard errors in the perimeter and area estimates for the two polygons. Results are listed in Table 7.2, where comparisons are made between the statistics obtained from the analytical and simulation methods.

Table 7.2 Error estimation for perimeters and areas based on
variance propagation and simulation (units: m^2).

	Objects							
	House (10 vertices)				Pond (136 vertices)			
	perimeter		area		perimeter		area	
	mean	std.	mean	std.	mean	std.	mean	std.
Calculated	48.1	2.5	119.0	8.0	479.7	9.6	7937.6	26.1
Simulated								
unconditional	48.9	2.1	123.2	8.4	504.6	2.8	8163.7	83.8
conditional	47.9	1.3	116.3	3.7	483.8	2.0	7923.8	34.5

As can be inferred from the data in Table 7.2, analytical methods for variance propagation in perimeter calculation report standard errors much greater than those derived from stochastic simulation, in particular, for the case of the pond. On the other hand, for area calculation, errors are underestimated by analytical approaches, although analytical estimation of error in area calculation for the house is much greater than that from conditionally simulated data, and only slightly smaller than that from unconditional simulation.

The overestimation for errors in perimeters is dominated by the fact that polygon boundaries are roughly, not smoothly, linked up from line segments. On the other hand, the underestimation of error in pond area arises from the recognition that the pond is shaped with a much smaller ratio of perimeter to area (approximately 6%) than for the house (about 40%), whose estimated error in area is biased towards over-estimation. In general, the overestimation of errors in perimeters on the one hand and underestimation of errors in areas by analytical approaches on the other are due to the assumption of spatial independence among positional errors in vertices defining polylines and polygon boundaries.

It can also be seen from Table 7.2 that conditional simulation tends to underestimate perimeters and areas, except for the case of pond perimeter, while unconditional simulation overestimates both perimeters and areas. Accordingly, significantly larger values of standard errors were obtained with unconditional than conditional methods of simulation.

As extra verification, using the χ^2 statistic with 9 degrees of freedom, null hypotheses of equal variances in perimeter and area estimates were rejected at a significance level of 5% for both polygons, confirming significant spatial correlation in positional errors. Of course, the existence of spatial dependence in positional errors has already been proved and modelled with semivariograms.

Thus, an objective assessment of the errors in area quantities computed from vector data of limited accuracy must be based on the appropriate accommodation of the spatial correlation intrinsic to positional errors; otherwise, error quantities may be surprisingly low or unacceptably high, casting doubt on common GIS functionality. Stochastic simulation provides a workable solution to the propagation of positional errors in line- and area-based spatial queries and applications.

7.5 DISCUSSION

This chapter has discussed the description and modelling of errors in object-based spatial databases, and examined the possible ways in which spatial dependence may be usefully explored for handling positional errors, the main concern in the domain of discrete objects. Stochastic simulation was favoured again over analytical approaches for its versatility in modelling positional errors and their propagation in lines and areas.

An empirical study using photogrammetric data based in an Edinburgh suburb has shown that geostatistics can be applied to analyse positional errors without major difficulties. As for field-based continuous and categorical data, spatial correlation that is found significant in positional errors can and should be incorporated into the analysis of object errors. Results reinforced the adaptability of stochastic simulation for mapping complexity in composite objects and their measurements. These methods allow error propagation from source data sets, such as polygon coverages to an overlaid coverage, to be evaluated flexibly. Based on error modelling, both data producers and data users are able to assess the fitness of a particular data product for a certain purpose.

Stochastic simulation is effectively carried out by drawing equal-probable realisations of underlying error-laden data from suitably designed computer programs operating on raster-based data structures. While the necessity for discrete vector representation of objects in spatial databases is widely admitted, the idea of using rasters of finite resolution for modelling uncertain vector data seems also to be growing in popularity. The arguments and experiments developed in this chapter provide a perspective on geographical data modelling, in which heterogeneous data of disparate formats and semantics may become interoperable in the sense that spatial errors are not impediments to integration, but phenomena to be explored.

For instance, as for field-based variables, positional errors in different object data sets may be processed in combination by using geostatistical approaches such as co-Kriging. Thus, theoretically sound techniques can be developed for map conflation, perhaps by weighting heterogeneous data in a way that is spatially adaptive and accuracy dependent. Moreover, the unified representations of positional errors and other originally field-modelled variables via raster format of varying resolutions may also facilitate geostatistical applications across the border of discrete objects and continuous fields, whereby spatial and cross-variable correlations contribute to interoperable processing of geographical data.

The experiment documented, following the theoretical exposition, considered a special type of vector data, comprising well-defined point and line objects, commonly known as vector data. This by-passed the issue of uncertainty in object identification and extraction, which is often predominant with raster data when such patterns are delineated from images. Uncertainty of object identification may be approachable with the availability of measurements and rules employed for object extraction based on remote-sensor images, which hinge increasingly on advanced methods for computerised feature extraction (Besag, 1986; Woodcock and Harward, 1992). Many questions remain in this area, although research efforts continue.

This chapter has focused on the case of vector-structured objects consisting of line segments. However, other types of vector data, such as those created from continuous curves, may be more powerful for object representation (Goodchild, 1998). As indicated by Figure 7.9, the semantics of lines behave differently between polylines and curves. Error models must be able to accommodate both for sensible interpretation. If lines and areas were simply aggregations of points, there would perhaps be little need for all the interest in spatial error handling, as the conventional least-squares method would have served GIS well enough. In fact, much of the strength of research into object uncertainties comes from a re-assessment of the formalism of points, lines, and areas. The approach proposed by Goodchild and Hunter (1997) and later pursued by Tveite and Langaas (1999) provides insights into the complexity that lies behind neatly and smoothly drawn objects, as discussed in this chapter.

A different approach is needed when individual points are all that are really required for establishing geodetic control. For the maintenance of precision positional networks, the coordinates computed from land surveying and other measurements are of little use, since data integrity and consistency might be affected in GIS updating. Instead, measurement-based GISs are created when original measurements of angles and distances and textual data are recorded in databases in lieu of representations of derived objects, allowing a consistent process of data updating (Durgin, 1993; Goodchild, 1999a). Such measurement-based GISs ensure that spatial objects are modified not from stored objects but from original measurements, which are kept up-to-date and are used for specific purposes.

The emphasis on positional errors in objects is in contrast to errors in continuous or categorical properties. For this reason, there is an accumulation of research efforts in the literature on applying novel methods such as geostatistics and fuzzy sets for enhanced geo-processing of spatially varied and correlated errors in qualitative and quantitative fields (Pawlowsky *et al.*, 1995; De Gruijter *et al.*, 1997). These methods have met with remarkable success in the analysis of remotely sensed images and other geographical data in raster format. It is thus clear that the experiment reported in this chapter extends the applicability of geostatistics to handling additional types of spatial error.

Many techniques developed for inherently raster-structured, field-modelled data could be considered for possible exploitation in the domain of objects. A special case for consideration is about poorly defined objects, such as boundaries in resources inventory data or informally named places, where boundary lines are actually zones of uncertainty and may not be interpreted on a purely geometric basis. This point has been highlighted by Burrough and Frank (1996), and addressed as part of the discussion of fuzzy methods for spatial uncertainty in Chapter 6. For intrinsically fuzzy things, the Boolean logic underlying conventional probabilistic methods is less suitable than fuzzy-set theory, and the semantics of fuzzy objects offer much flexibility (Cheng *et al.*, 2001).

Object models for geographical abstraction and data representation are conceptually different from fields. Such a distinction, although usually necessary in theoretical terms, should not be pushed to extremes where the two perspectives for geographical abstraction are locked in isolation, losing sight of field and object characteristics that are complementary to each other. A valuable modification to stochastic simulation would be to enforce object rigidity and shape regularities such as rectangles and parallels in distorted realisations of objects, and to tune simulated fields to honour discontinuities. As the real world is complex and rarely fits neatly into the stereotypes of either continuous fields or discrete objects alone, it is usually a combination of field and object conceptualisations that is really involved in most practical settings, with continuous fields and discrete objects representing the two extremes. To some extent, the uncertainties that are endemic in GIS sustain the continuum view of fields and objects, and blur their methodological boundaries.

A goal for uncertainty handling in general is to reduce uncertainty in GIS data, and ultimately to enhance the reliability and predictability of spatial problem-solving and decision-making. Towards this aim, techniques specifically tailored for error handling in continuous fields and discrete objects might be integrated under the banner of stochastic simulation. Developments in techniques such as GIS interoperability and data conflation have much to offer in making sensible use of the increasingly vast resources of spatial data of varied lineage.

CHAPTER 8

Uncertainty-Informed Geographies

8.1 GEOMATICS OF UNCERTAINTY

Geographical information has played an important role in exploration, and has formed an essential part of human civilisation's cultural and scientific resources. Geographical studies were prominent activities in the age of renaissance, and at the forefront of globalisation in that term's widest possible sense. Today, much of the excitement of early European exploration is captured by programs of planetary exploration, and by the scientific exploration of the oceans. Modern societies are conscious of the need for better geographical information related to such issues as biodiversity and sustainable development. Global trading is also increasing the value of geographical information, and so too is interest in localisation, and the preservation of indigenous ways of life. Many common tasks of way-finding, site selection, resource management, and spatial optimisation have been extensively computerised, and rely on abundant supplies of current and accurate geographical information. Analysis of such information can now be performed without many of the old constraints of space and time.

GIS, remote sensing, mobile mapping, and geo-computation are major digital technologies supporting the processing and communication of geographical information. These cutting-edge technologies have developed to a point where a strong scientific and conceptual basis is required. Uncertainty is an inseparable component in science, and information and understanding about uncertainty are of crucial importance for the scientific development of geographical information technologies. The long-term effects of enhanced uncertainty handling in GISs and their peripheral technologies will be of both economic and cultural benefit, if they can be based on enlightened approaches to the collection, representation, analysis, and communication of scientific information about geographical uncertainty.

In a digital world, geographical distributions and entities are abstracted appropriately as either fields or objects, subject to real-world limitations, modelling requirements, data availability, and technical constraints. Fields are discretised into locations associated with specific qualities or quantities, while objects are positioned with coordinates, and attributed with alpha-numeric descriptions following correct identification. Field-based models are naturally favoured in physical geography and environmental applications, and have shown remarkable flexibility in geographical modelling coupled with stochastic constructs of spatial variability and correlation. Object-based models are more suitable for cartography and facility management, because geometric calculation, inference about topologic relations, and object-based spatial reasoning are well addressed by the metaphors of discrete objects.

The process of geographical abstraction that creates the two alternative data models imposes generalisation, selection, and approximation prior to subsequent information processing. Fields may be represented in the format of location-specific samples. Methods of interpolation over space are not standardised, and multiple fields may be created from a single data set. The plurality of data models suitable for field-based representation complicates uncertainty in fields further. More notorious is the case of discrete objects, which seldom exist in the real world. Even engineered spatial structures, such as major highways, are often simplified into polylines to allow for easy computing and mapping.

Geographical data, as raw components of geographical models, are reduced from various measurements, which are obtained through complex interactions of human subjectivity and the physics of sensors. Thus, data recorded in spatial databases are of finite resolution and limited accuracy, and they pertain only to partial variations of underlying geographies. Therefore, geographical investigation and information processing must be conducted with the understanding that the data sets that have been collected and stored are simplified and smoothed segments of more pronounced spatial variability and heterogeneity. In other words, real spatially distributed phenomena are generalised, selected, and approximated with certain thresholds and tolerances that may or may not be stated explicitly. Maps that depict geographical distributions are subject to a host of unknown factors governing spatial modelling, measurement, and cartographic production. Thus, digitised maps do not diminish but actually accentuate the inherent inaccuracy in geographical modelling and mapping.

With geographical models and measurements suffering approximation and inaccuracy, geographical information can no longer be considered as error-free tuples recorded in spatial databases. What are actually stored in GIS are realised versions (x',G') drawn from stochastic distributions of tuples (x,G), which are often unknown or only known to a certain extent. Because geographical data recorded in GIS are only realisations of the underlying phenomena, they can only approximate the truth.

Uncertainty should be more objectively viewed as akin to spatially distributed tuples (x',G') than to difference tuples $(x-x',G-G')$, as the latter are generally not attainable due to difficulties in defining the truth tuples (x,G). This assertion has a significant bearing on uncertainty modelling with respect to geographical phenomena, since it implies that uncertainty should be considered as capturing the information richness beneath the nominal contents of data sets. In other words, uncertainty should be conceived of as an integral part of the representation, analysis, and investigation of geographical reality so that geographical information is provided and used with well-informed understanding of uncertainties.

While it may be difficult to conduct repeated measurement at every location in order to measure uncertainty in geographical variables directly, the notions of distribution stationarity and spatial dependence offer clues to the complete reconstruction of spatial variability from limited samples. Thus, the operational implementation of uncertainty modelling is no longer as remote as one might have assumed.

For fields, uncertainty modelling seems to be appropriately addressed by stochastic methods. For some, it may be all that is needed to support quantification of spatial errors. For others, it may be more interesting to formulate probabilistic statements about localised occurrences of errors or possible structural patterns in uncertain distributions. Kriging and indicator Kriging provide extended applications of geostatistics for mapping uncertainty. More versatile mechanisms are supported by stochastic simulation, which can be designed to generate equally probable realisations of spatial variables. Simulated fields are fed into spatial functions to permit analysis of error propagation, even though the spatial functions may be non-differentiable, non-numeric, and simply arbitrary.

Variables may be defined and measured on continuous or categorical scales. Spatial statistics offer an appropriate way of representing uncertain behaviours in spatial quantities by using randomised numbers that catch the stochastics of the underlying phenomena. Any spatial structures relevant to the underlying fields can be used to constrain the realisations through the enforcement of spatial correlations inferred from sampled data or asserted by reasonable deduction. For handling uncertain spatial categories, an enabling mechanism is the indicator transform, by which categories prevailing at individual locations are encoded as binary data corresponding to either a confirmed or a refuted class membership. The transcendent nature of probabilities and probabilistic interpretation lead equally well to the flexible description and modelling of categorical occurrences over space.

Fields and objects function distinctively. To implement stochastic simulation of errors in objects, one has somehow to overcome the issues of spatial discreteness central to object-based geographical abstraction. As position and attributes are separable, or assumed so for well-defined discrete objects, the continuity of spatial coordinates suggests a viable extension of spatial statistics to modelling positional errors. Apparently rigid objects can be distorted elastically through the modelling of positional displacement. In the field domain, errors are high-pass versions of vector distortion fields, and may be seen as de-trended. Positional errors function in a similar way. As special types of fields, positional errors can be analysed using concepts of spatial continuity and dependence. Uncertainty in lines and areas can be treated in a way that is easier and more tractable based on point-based models, especially in the case of complex objects where analytical estimation seems impractical.

A useful assumption underlying stochastic modelling is stationarity, which is implicitly made for models rather than for the underlying reality. Much of geostatistics is built on stationarity and smoothed models of spatial dependence. Gaussian models are often utilised, and are widely acclaimed by modellers as suitable tools for data exploration. But other non-parametric solutions are required in the presence of spatial non-stationarity and abnormality. For this, a wealth of stochastic models may be usefully incorporated for solving spatial problems.

In the family of spatial statistics, a notable family of methods for stochastic modelling is built upon the Markov specification of local conditions for stochastic distributions. Makovian mathematics has a matching counterpart in physics, in areas such as Gibbs mechanics. Many different forms of energy functions can be invoked as the basis for the interpretation of mechanisms of uncertainty-prone processes. Algorithms that are well tested can be adopted with objective functions that reflect the specifics of a particular system. Concepts of space and neighbourhood may become non-Euclidean to allow for geographical discontinuity and irregularity. In stochastic problem-solving, Bayesian strategies clearly offer viable approaches by providing a unique way for updating information in the light of new data. Both prior and posterior distributions may benefit from Markov models of local interactions of fields and objects.

In the light of these diverse conceptual and modelling frameworks, it is clear that data remain more reliable than models, even though the latter are steadily evolving. Geographical information does not exist in a vacuum of idealised models, but in a complex environment that frequently appears chaotic. Geographical information and knowledge discovery are meaningful only to the extent that data and analysis are accurate. As geographical inquiries generate fresh insights and update human understanding about our surroundings, there will continue to be a complex tension between geographical certainty and geographical uncertainty.

In summary, not only are stochastic simulators able to produce sequences of sampled fields conforming to conditioning data and known spatial characteristics, but they can also provide ways of solving problems that would otherwise be insoluble with deterministic mathematics. The most important contribution of the stochastic paradigm to geographical computing and thinking is that it fosters a closer link between geographical phenomena and computational complexity. This assertion is clearly demonstrated by empirical studies with real data sets in practical settings.

Geographical information scientists are making substantial progress in processing and communicating uncertainties. But two outstanding and apparently conflicting issues stand in the way. One is the need for methodological integration of geographical modelling and uncertainty computation, and the other is the need for less mathematical and operational complexity in pursuing uncertainty handling, without at the same time undermining the objectivity of uncertainty modelling. The former aims for efficiency and enhanced overall computability of geographical uncertainties, along with accuracy in uncertainty reporting. Here, the potential for integration of fields and objects in the context of uncertainties presents a stimulating topic for further research. In the latter case, the solution is clearly to simplify representational and stochastic details, which could be complicated enough to deter potentially interested parties from exploring these unfamiliar territories. Scientists should work to assist people to overcome the bewildering aspects of mathematical and statistical analysis, and geographical information scientists are no exception.

8.2 AWARENESS OF UNCERTAINTY

While statistical estimation, probabilistic inference, and stochastic simulation have shown their value in dealing with spatial uncertainty, other forms of uncertainty are perhaps better studied outside the framework of probability. For instance, daily discourse concerning geographical phenomena is frequently full of vagueness. To most people, coastlines are vaguely defined as geometric demarcations between land and sea. The Scottish highlands have equally vague spatial boundaries and cultural meanings. Torrential rains, floods, and high tides are not exactly well defined either. Property assessment must juggle many complex factors, many of which are not easy to categorise. Speculation over the volume of an oil reserve is often underlain by vagueness in the definition of the reserve. All such vague geographies are better treated as fuzzy sets rather than as crisp sets. And in the face of fuzziness, Boolean logic is surely less versatile than fuzzy logic in dealing with discourse that is full of heuristic metaphors, linguistic hedges, and other forms of subjectivity.

Vagueness seems more prevalent in geographical information than in other forms of information and domains of applications. This may be reasonable if one recognises that geographers must somehow deal with the vast spatial, taxonomic, and temporal complexity of geographical reality. While samples of soils may be dug up and subjected to intensive analysis in the laboratory, the great majority of geographically distributed phenomena must be sampled directly in the field, with constraints of labour and other expenses. Even if one were able to stage a careful and comprehensive investigation at a single site, there would still be enough fuzziness to render such an endeavour frustrating, if not fruitless. Vagueness seems inevitable in representing geographical reality, and humans must seek ways of encoding such vague information to permit computation, automated reasoning, and sharing.

While frequentist and subjective probabilities appear to be incomplete as conceptualisations of various kinds of vagueness, fuzzy sets appear better able to mimic human expression and reasoning. The schools of probabilistic and fuzzy evaluation seem complementary to each other as bases for analysing disjoint uncertainties of occurrences, degrees of confirmation, and strengths of belonging.

However, both theories seem applicable in some settings but inapplicable in others. The theoretical soundness of probability is not well matched with practical tangibility, as it may be impossible to conduct random sampling for an empirical assessment. Subjective probabilities are even more questionable, as they are often assessed without clear metaphors of idealised events and sample spaces. Fuzzy sets are clearly defined in a weak framework, and are poor in semantics. Thus, the methodological competitiveness of probability and fuzzy-set theory may present a serious dilemma to some. But although purists may be uneasy with what are clearly theoretical and practical discrepancies, the need for flexibility suggests that probabilistic and fuzzy methods should be pursued in combination.

In addition to enhanced capabilities in spatial reasoning with fuzzy logic, visual techniques are also valuable ways of investigating geographical phenomena. Maps developed historically as more powerful media than text for geographical information. GIS and geographical computing should not diminish but enrich this tradition. It has been shown that visualisation can be used to enhance the communication of uncertain data in spatial analysis and decision-making, leading to improved knowledge about underlying reality. Uncertainties may exhibit certain spatial patterns, and visualisation can help by exposing the occurrences and behaviours of geographical uncertainties.

Visual computing may well shed light on uncertainty and its role in spatial queries and analysis with GIS. The current implementation of uncertainty handling in GIS is sufficiently shrouded with mathematical details to remain unpopular, and its use is effectively limited to experts with strong grounding in mathematics and statistics. Even with more accessible computational systems for handling geographical uncertainties, the complex nature of databases and algorithms tends to discourage collaborative interaction between human and machine. Uncertainty in geographical databases complicates computing further. The use of visual metaphors instead of unnecessary textual description can provide the basis for more friendly user interfaces during geographical computing. This way, uncertainty modelling can become more accessible to the communities of users of geographical information, while remaining a serious business. Once uncertainties are encapsulated with data sets and modelled via stochastic and fuzzy computing, visual methods can be invoked for analysing such data sets, and for assessing their fitness of use in specific applications.

The perspective on uncertainty handling presented in this book has been built upon the dual concepts of fields and objects, and has drawn on the theories and methods of spatial statistics, probability theory, and fuzzy sets. The ultimate goal of uncertainty modelling has to be the formulation of strategies for error management, that is, for decision-making in the presence of errors, and strategies for the reduction or elimination of error in output products. These two interconnected problems are driven by the inferences that may be drawn from the results of error propagation. This process is very crucial, as it serves to justify all the investment and effort directed towards uncertainty handling. Systems for uncertainty modelling cannot be locked up in a cupboard, but must be brought close to ordinary lives, through a process of communication, and towards the objective of improved geographical information services and productivity. Academics, technologists, government information agencies, the general public, and the commercial sector must work together to take advantage of the benefits of geographical information in new applications, while being fully informed of the nature and implications of the associated uncertainties. Scientists and workers lead the leap forward!

References

Abeyta, A.M. and Franklin, J., 1998, The accuracy of vegetation stand boundaries derived from image segmentation in a desert environment. *Photogrammetric Engineering and Remote Sensing*, **64**, pp. 59–66.

Abler, R.F., 1987, The National Science Foundation Center for Geographic Information and Analysis. *International Journal of Geographical Information Systems*, **1**, pp. 303–326.

Ahlqvist, O., Keukelaar, J. and Oukbir, K., 2000, Rough classification and accuracy assessment. *International Journal of Geographical Information Science*, **14**, pp. 475–496.

Almeida, A.S. and Journel, A.G., 1994, Joint simulation of multiple variables with a Markov-type coregionalization model. *Mathematical Geology*, **26**, pp. 565–588.

American Society for Photogrammetry and Remote Sensing (ASPRS), 1989, ASPRS interim accuracy standards for large scale line maps. *Photogrammetric Engineering and Remote Sensing*, **55**, pp. 1038–1040.

Anderson, J.E., Hardy, E.E., Roach, J.T. and Witmer, R.E., 1976, A land use and land cover classification system for use with remote-sensor data. *U.S. Geological Survey Professional Paper 964*.

Anselin, L., 1993, Discrete space autoregressive models. In *Environmental Modeling with GIS*, edited by Goodchild, M.F., Parks, B. and Steyaert, L. (New York: Oxford University Press), pp. 454–469.

Arbia, G., Griffith, D. and Haining, R., 1999, Error propagation modelling in raster GIS: adding and ratioing operations. *Cartography and Geographic Information Science*, **26**, pp. 297–315.

Aronoff, S., 1993, *Geographic Information Systems: a Management Perspective* (Ottawa: WDL Publications).

Atkinson, P.M. and Tate, N.J., 2000, Spatial scale problems and geostatistical solutions: a review. *Professional Geographers*, **52**, pp. 607-623.

Aumann, G., Ebner, H. and Tang, L., 1991, Automatic derivation of skeleton line from digitized contours. *ISPRS Journal of Photogrammetry and Remote Sensing*, **46**, pp. 259–268.

Bailey, R.G., 1988, Problems with using overlay mapping for planning and their implications for geographic information systems. *Environmental Management*, **12**, pp. 11–17.

Baillard, C. and Dissard, O., 2000, A stereo matching algorithm for urban digital elevation models. *Photogrammetric Engineering and Remote Sensing*, **66**, pp. 1119–1128.

Balce, A.E., 1987, Determination of optimum sampling interval in grid digital elevation models (DEM) data acquisition. *Photogrammetric Engineering and Remote Sensing*, **53**, pp. 323–330.

Banai, R., 1993, Fuzziness in geographical information systems: contributions from the analytic hierarchy process. *International Journal of Geographical*

Information Systems, **7**, pp. 315–329.

Barnsley, M.J. and Barr, S.L., 1996, Inferring urban land use from satellite sensor images using kernel-based spatial reclassification. *Photogrammetric Engineering and Remote Sensing*, **62**, pp. 949–958.

Barrett, J.P. and Philbrook, J.S., 1970, Dot grid area estimates: precision by repeated trials. *Journal of Forestry*, **68**, pp. 149–151.

Bartlett, M.S., 1967, Inference and stochastic processes. *Journal of the Royal Statistical Society*, series A, **130**, pp. 459–477.

Bauer, M.E., Hixson, M.M., Davis, B.J. and Etheridge, J.B., 1978, Area estimation of crops by digital analysis of Landsat data. *Photogrammetric Engineering and Remote Sensing*, **44**, pp. 1033–1043.

Berry, J.K., 1987, Computer-assisted map analysis: potential and pitfalls. *Photogrammetric Engineering and Remote Sensing*, **53**, pp. 1405–1410.

Besag, J.E., 1974, Spatial interaction and the statistical analysis of lattice schemes. *Journal of the Royal Statistical Society*, series B, **36**, pp. 192–225; Discussion, pp. 225–236.

Besag, J.E., 1986, On the statistical analysis of dirty pictures. *Journal of the Royal Statistical Society*, Series B, **48**, pp. 259–279; Discussion, pp. 280–302.

Bezdek, J.C., Ehrlich, R. and Full, W., 1984, FCM: the fuzzy *c*-means clustering algorithm. *Computers and Geosciences*, **10**, pp. 191–203.

Bierkens, M. and Burrough, P., 1993, The indicator approach to categorical soil data. I. Theory. *Journal of Soil Science*, **44**, pp. 361–368.

Binaghi, E., Brivio, P.A., Ghezzi, P. and Rampini, A., 1999, A fuzzy set-based accuracy assessment of soft classification. *Pattern Recognition Letters*, **20**, pp. 935–948.

Black, M., 1937, Vagueness: an exercise in logical analysis. *Philosophy of Science*, **4**, pp. 427–455.

Blakemore, M., 1984, Generalisation and error in spatial databases. *Cartographica*, **21**, pp. 131–139.

Bolstad, P., Gessler, P., and Lillesand, T.M., 1990, Positional uncertainty in manually digitized map data. *International Journal of Geographical Information Systems*, **4**, pp. 399–412.

Bracken, I. and Martin, D., 1989, The generation of spatial population distributions from census centroid data. *Environment and Planning* A, **21**, pp. 537–44.

Brassel, K.E. and Weibel, R., 1988, A review and conceptual framework of automated map generalization. *International Journal of Geographical Information Systems*, **2**, pp. 229–244.

Brémaud, P., 1999, *Markov Chains: Gibbs Fields, Monte Carlo Simulation, and Queues* (New York: Springer-Verlag).

Bresenham, J.E., 1965, Algorithm for computer control of a digital plotter. *IBM Systems Journal*, **4**, pp. 25–30.

Brown, D.G., 1998, Classification and boundary vagueness in mapping presettlement forest types. *International Journal of Geographical Information Science*, **12**, pp. 105–129.

Burkholder, E.F., 1999, Spatial data accuracy as defined by the GSDM. *Surveying and Land Information Systems*, **59**, pp. 26–30.

Burrough, P.A. and Frank, A.U. (Eds), 1996, *Geographic Objects with Indeterminate Boundaries* (Basingstoke: Taylor & Francis).

Burrough, P.A. and McDonnell, R.A., 1998, *Principles of Geographical Information Systems* (Oxford and New York: Oxford University Press).

Campbell, J.B., 1981, Spatial correlation effects upon accuracy of supervised classification of land cover. *Photogrammetric Engineering and Remote Sensing*, **46**, pp. 353–363.

Campbell, J.B., 1996, *Introduction to Remote Sensing*, 2nd edition (New York and London: The Guilford Press).

Card, D.H., 1982, Using known map category marginal frequencies to improve estimates of thematic map accuracy. *Photogrammetric Engineering and Remote Sensing*, **48**, pp. 431–439.

Carnahan, W.H. and Zhou, G., 1986, Fourier transform techniques for the evaluation of the Thematic Mapper line spread function. *Photogrammetric Engineering and Remote Sensing*, **52**, pp. 639–648.

Carr, J.R., 1996, Spectral and textural classification of single and multiple band digital images. *Computers and Geosciences*, **22**, pp. 849–865.

Carter, J.R., 1988, Digital representation of topographic surfaces. *Photogrammetric Engineering and Remote Sensing*, **54**, pp. 1577–1580.

Carter, J.R., 1992, The effect of data precision on the calculation of slope and aspect using gridded DEMs. *Cartographica*, **29**, pp. 22–34.

Cheng, T., Molenaar, M. and Lin, H., 2001, Formalizing fuzzy objects from uncertain classification results. *International Journal of Geographical Information Science*, **15**, pp. 27–42.

Chorley, R.J. and Haggett, P. (Eds), 1968, *Models in Geography* (London: Methuen).

Chrisman, N.R., 1989, Modelling error in overlaid categorical maps. In *Accuracy of Spatial Databases*, edited by Goodchild, M.F. and Gopal, S. (Basingstoke: Taylor and Francis), pp. 21–34.

Chrisman, N.R., 1991, The error component in spatial data. In *Geographical Information Systems: Principles and Applications*, edited by Maguire, D.J., Goodchild, M.F. and Rhind, D.W. (Harlow: Longman Scientific and Technical), Vol. 1, pp. 165–174.

Chrisman, N.R. and Yandell, B.S., 1988, Effects of point error on area calculations: a statistical model. *Surveying and Mapping*, **48**, pp. 241–246.

Clarke, K.C., 1995, *Analytical and Computer Cartography*, 2nd edition, (Englewood Cliffs, N.J.: Prentice Hall).

Cliff, A.D. and Ord, J.K., 1973, *Spatial Autocorrelation* (London: Pion).

Coad, P., North, D. and Mayfield, M., 1995, *Object Models: Strategies, Patterns, and Applications* (Englewood Cliffs, NJ: Prentice Hall).

Cohen, J., 1960, A coefficient of agreement for nominal scales. *Educational and Psychological Measurement*, **20**, pp. 37–46.

Congalton, R.G., 1991, A review of assessing the accuracy of classifications of remotely sensed data. *Remote Sensing of Environment*, **37**, pp. 35–46.

Congalton, R.G. and Green, K., 1999, *Assessing the Accuracy of Remotely Sensed Data: Principles and Practices* (Boca Raton: CRC Press, Inc.).

Cooper, M.A.R. and Cross, P.A., 1988, Statistical concepts and their application in photogrammetry and surveying. *Photogrammetric Record*, **12**, pp. 637–663.

Coppock, J.T. and Rhind, D.W., 1991, The history of GIS. In *Geographical Information Systems: Principles and Applications*, edited by Maguire, D.J., Goodchild, M.F. and Rhind, D.W. (Harlow: Longman Scientific and Technical), Vol. 1, pp. 21–43.

Couclelis, H., 1999, Space, time, geography. In *Geographical Information Systems: Principles, Techniques, Applications and Management*, edited by Longley, P.A., Goodchild, M.F., Maguire, D.J. and Rhind, D.W. (New York: John Wiley & Sons, Inc.), pp. 29–38.

Cracknell, A.P., 1998, Synergy in remote sensing—what's in a pixel? *International Journal of Remote Sensing*, **19**, pp. 2025–2047.

Crapper, P.F., 1980, Errors incurred in estimating an area of uniform land cover using Landsat. *Photogrammetric Engineering and Remote Sensing*, **46**, pp. 1295–1301.

Crapper, P.F., 1984, An estimate of the number of boundary cells in a mapped landscape coded to grid cells. *Photogrammetric Engineering and Remote Sensing*, **50**, pp. 1497–1503.

Cressie, N.A.C., 1993, *Statistics for Spatial Data,* revised edition (New York: John Wiley & Sons, Inc.).

Curlander, J.C. and McDonough, R.N., 1991, *Synthetic Aperture Radar: Systems and Signal Processing* (New York: John Wiley & Sons, Inc.).

Curran, P.J., 1988, The semi-variogram in remote sensing: an introduction. *Remote Sensing of Environment*, **24**, pp. 493–507.

Davis, F.W. and Simonett, D.S., 1991, GIS and remote sensing, In *Geographical Information Systems: Principles and Applications*, edited by Maguire, D.J., Goodchild, M.F. and Rhind, D.W. (Harlow: Longman Scientific and Technical), Vol. 1, pp. 191–213.

De Bruin, S., 2000a, Predicting the areal extent of land-cover types using classified imagery and geostatistics. *Remote Sensing of Environment*, **74**, pp. 387–396.

De Bruin, S., 2000b, Querying probabilistic land cover data using fuzzy set theory. *International Journal of Geographical Information Science*, **14**, pp. 359–372.

DeFries, R.S. and Townshend, J.R.G., 1994, NDVI-derived land cover classifications at a global scale. *International Journal of Remote Sensing*, **15**, pp. 3567–3586.

De Groeve, T., Lowell, K. and Thomson, K., 1999, Super ground truth as a foundation for a model to represent and handle spatial uncertainty. In *Spatial Accuracy Assessment: Land Information Uncertainty in Natural Resources*, edited by Lowell, K.E. and Jaton, A. (Chelsea: Ann Arbor Press), pp. 189–193.

De Gruijter, J.J., Walvoort, D.J.J. and van Gaans, P.F.M., 1997, Continuous soil maps—a fuzzy set approach to bridge the gap between aggregation levels of process and distribution models. *Geoderma*, **77**, pp. 169–195.

Delfiner, P. and Delhomme, J.P., 1975, Optimum interpolation by Kriging. In *Display and Analysis of Spatial Data*, edited by Davis, J.C. and McCullagh, M.J. (London and New York: John Wiley & Sons, Inc.), pp. 96–114.

Dempster, A.P., 1967, Upper and lower probabilities induced by a multivalued mapping. *Annals of Mathematical Statistics*, **38**, pp. 325–339.

Deutsch, C.V. and Journel, A.G., 1998, *GSLIB: Geostatistical Software Library and User's Guide*, 2nd edition (New York and Oxford: Oxford University Press).

De Vries, P.G., 1986, *Sampling Theory for Forest Inventory: a Teach-Yourself Course* (Berlin: Springer-Verlag).

Diggle, P.J., Tawn, J.A. and Moyeed, R.A., 1998, Model-based geostatistics. *Applied Statistics*, **47** (part 3), pp. 299–350.

Douglas, D.H. and Peucker, T.K., 1973, Algorithms for the reduction of the number of points required to represent a digitized line or its caricature. *The Canadian Cartographer*, **10**, pp. 112–122.

Dozier, J., 1981, A method for satellite identification of surface temperature fields of sub-pixel resolution. *Remote Sensing of Environment*, **11**, pp. 221–229.

Drummond, J.E., 1995, Positional accuracy. In *Elements of Spatial Data Quality*, edited by Guptill, S.C. and Morrison, J.L. (Oxford: Elsevier Scientific), pp. 31–58.

Dubois, D. and Prade, H., 1997, The three semantics of fuzzy sets. *Fuzzy Sets and Systems*, **90**, pp. 141–150.

Duckham, M., Mason, K., Stell, J. and Worboys, M., 2001, A formal approach to imperfection in geographic information. *Computers, Environment and Urban Systems*, **25**, pp. 89–103.

Duda, R.O. and Hart, P.E., 1973, *Pattern Classification and Scene Analysis* (New York: John Wiley& Sons, Inc.).

Duggin, M.J. and Robinove, C.J., 1990, Assumptions implicit in remote sensing data acquisition and analysis. *International Journal of Remote Sensing*, **11**, pp. 1669–1694.

Dungan, J., 1998, Spatial prediction of vegetation quantities using ground and image data. *International Journal of Remote Sensing*, **19**, pp. 267–285.

Dunn, R., Harrison, A.R. and White, J.C., 1990, Positional accuracy and measurement error in digital databases of land use: an empirical study. *International Journal of Geographical Information Systems*, **4**, pp. 385–398.

Durgin, P.M., 1993, Measurement-based databases: one approach to the integration of survey and GIS cadastral data. *Surveying and Land Information Systems*, **53**, pp. 41–47.

Dutton, G., 1992, Handling positional uncertainty in spatial databases. In *Proceedings of 5th International Symposium on Spatial Data Handling*, pp. 460–469.

Edwards, G. and Lowell, K.E., 1996, Modelling uncertainty in photointerpreted boundaries. *Photogrammetric Engineering and Remote Sensing*, **62**, pp. 373–391.

Ehlers, M., 1985, The effects of image noise on digital correlation probability.

Photogrammetric Engineering and Remote Sensing, **51**, pp. 357–365.

Englund, E., 1993, Spatial simulation: environmental applications. In *Environmental Modeling with GIS*, edited by Goodchild, M.F., Parks, B. and Steyaert, L. (New York: Oxford University Press), pp. 432–437.

Estes, W.K., 1994, *Classification and Cognition* (New York and Oxford: Oxford University Press).

Ferns, D.C., Zara, S.J. and Barber, J., 1984, Application of high resolution spectroradiometry to vegetation. *Photogrammetric Engineering and Remote Sensing*, **50**, pp. 1725–1735.

Filin, S. and Doytsher, Y., 1999, A linear mapping approach to map conflation: matching of polylines. *Surveying and Land Information Systems*, **59**, pp. 107–114.

Fine, T.L, 1973, *Theories of Probability: an Examination of Foundations* (New York: Academic Press).

Fishburn, P.C., 1986, The axioms of subjective probability. *Statistical Science*, **1**, pp. 335–345; Discussion, pp. 346–358.

Fisher, P.F., 1991, Modelling soil map-unit inclusions by Monte Carlo simulation. *International Journal of Geographical Information Systems*, **5**, pp. 193–208.

Fisher, P.F., 1998, Improved modelling of elevation error with geostatistics. *GeoInformatica*, **2**, pp. 215–233.

Fisher, P.F., 2000, Sorites paradox and vague geographies. *Fuzzy Sets and Systems*, **113**, pp. 7–18.

Florinsky, I.V., 1998, Accuracy of local topographic variables derived from digital elevation models. *International Journal of Geographical Information Science*, **12**, pp. 47– 62.

Foody, G.M., 1995, Land cover classification by an artificial neural network with ancillary information. *International Journal of Geographical Information Systems*, **9**, pp. 527–542.

Foody, G.M., 1996, Approaches for the production and evaluation of fuzzy land cover classifications from remotely-sensed data. *International Journal of Remote Sensing*, **17**, pp. 1317–1340.

Foody, G.M., Campbell, N.A., Trodd, N.M. and Wood, T.F., 1992, Derivation and applications of probabilistic measures of class membership from the maximum likelihood classification. *Photogrammetric Engineering and Remote Sensing*, **58**, pp. 1335–1341.

Forster, B., 1983, Some urban measurements from Landsat data. *Photogrammetric Engineering and Remote Sensing*, **49**, pp. 1693–1707.

Franklin, J. and Woodcock, C.E., 1997, Multiscale vegetation data for the mountains of Southern California: spatial and categorical resolution. In *Scale in Remote Sensing and GIS*, edited by Quattrochi, D.A. and Goodchild, M.F. (Boca Raton: CRC/Lewis Publishers Inc.), pp. 141–168.

Fraser, C.S. and Shortis, M.R., 1992, Variation of distortion within the photographic field. *Photogrammetric Engineering and Remote Sensing*, **58**, pp. 851–855.

Frederiksen, P., Jacobi, O. and Kubik, K., 1985, A review of current trends in terrain modelling. *I.T.C. Journal*, **2**, pp. 101–106.

Freund, J.E., 1973, *Introduction to Probability* (Encino, CA: Dickenson Publishing Company, Inc.).

Frolov, Y.S. and Maling, D.H., 1969, The accuracy of area measurements by point counting techniques. *The Cartographic Journal*, **6**, pp. 21–35.

Fuller, R.M., Wyatt, B.K. and Barr, C.J., 1998, Countryside survey from ground and space: different perspectives, complementary results. *Journal of Environmental Management*, **54**, pp. 101–126.

Funk, C., Curtin, K., Goodchild, M.F., Montello, D. and Noronha, V., 1999, Formulation and test of a model of positional distortion fields. In *Spatial Accuracy Assessment: Land Information Uncertainty in Natural Resources*, edited by Lowell, K.E. and Jaton, A. (Chelsea: Ann Arbor Press), pp. 131–137.

Gahegan, M. and Ehlers, M., 2000, A framework for the modelling of uncertainty between remote sensing and geographic information systems. *ISPRS Journal of Photogrammetry and Remote Sensing*, **55**, pp. 176–188.

Garcia, J.A., Fdez-Valdivia, J. and Perez de la Blanca, N., 1995, An autoregressive curvature model for describing cartographic boundaries. *Computers and Geosciences*, **21**, pp. 397–408.

Geman, S. and Geman, D., 1984, Stochastic relaxation, Gibbs distribution, and the Bayesian restoration of images. *IEEE Transactions in Pattern Analysis and Machine Intelligence*, **6**, pp. 721–741.

Giles, P.T. and Franklin, S.E., 1998, An automated approach to the classification of the slope units using digital data. *Geomorphology*, **21**, pp. 251–265.

Good, I.J., 1950, *Probability and the Weighting of Evidence* (London: Charles Griffin & Company Ltd.).

Goodchild, M.F., 1978, Statistical aspects of the polygon overlay problem. In *Harvard Papers on Geographic Information Systems*, edited by Dutton, G. (Reading, MA: Addison-Wesley).

Goodchild, M.F., 1980, Algorithm 9: simulation of autocorrelation for aggregate data. *Environment and Planning* A, **12**, pp. 1073–1081.

Goodchild, M.F., 1988, The issue of accuracy in global databases. In *Building Databases for Global Science*, edited by Mounsey, H. and Tomlinson, R. (London: Taylor and Francis), pp. 31–48.

Goodchild, M.F., 1989, Modelling error in objects and fields. In *Accuracy of Spatial Databases*, edited by Goodchild, M.F. and Gopal, S. (Basingstoke: Taylor & Francis), pp. 107–113.

Goodchild, M.F., 1991, Issues of quality and uncertainty. In *Advances in Cartography*, edited by Müller, J.C. (London and New York: Elsevier Science Publication Ltd.), pp. 113–139.

Goodchild, M.F., 1992, Geographical information science. *International Journal of Geographical Information Systems*, **6**, pp. 31–46.

Goodchild, M.F., 1993, The state of GIS for environmental problem-solving. In *Environmental Modeling with GIS*, edited by Goodchild, M.F., Parks, B.O. and Steyaert, L.T. (New York and Oxford: Oxford University Press), pp. 9–15.

Goodchild, M.F., 1994, Integrating GIS and remote sensing for vegetation analysis and modeling: methodological issues. *Journal of Vegetation Science*, **5**, pp. 615–626.

Goodchild, M.F., 1995, Attribute accuracy. In *Elements of Spatial Data Quality*, edited by Guptill, S.C. and Morrison, J.L. (Oxford: Elsevier Scientific), pp. 59–79.

Goodchild, M.F., 1996, The spatial data infrastructure of environmental modeling. In *GIS and Environmental Modeling: Progress and Research Issues*, edited by Goodchild, M.F., Steyaert, L.T., Parks, B.O., Johnston, C.A., Maidment, D.R., Crane, M.P. and Glendinning, S. (Fort Collins, CO: GIS World Books), pp. 11–16.

Goodchild, M.F., 1998, Different data sources and diverse data structures: metadata and other solutions. In *Geocomputation: A Primer*, edited by Longley, P.A., Brooks, S.M., McDonnell, R. and Macmillan, W. (London: John Wiley & Sons, Inc.), pp. 61–74.

Goodchild, M.F., 1999a, Keynote speech: Measurement-based GIS. In *Proceedings of The International Symposium on Spatial Data Quality*, edited by Shi, W., Goodchild, M.F. and Fisher, P.F. (Hong Kong: Hong Kong Polytechnic University), pp. 1–9.

Goodchild, M.F., 1999b, Implementing Digital Earth: a research agenda. In *Towards Digital Earth: Proceedings of the International Symposium on Digital Earth*, edited by Xu, G. and Chen, Y. (Beijing: Science Press), Vol. 1, pp. 21–26.

Goodchild, M.F., Cova, T.J. and Ehlschlaeger, C.R., 1995, Mean objects: extending the concept of central tendency to complex spatial objects in GIS. In *Proceedings GIS/LIS '95*, Nashville, TN, pp. 354–364.

Goodchild, M.F., Egenhofer, M.J., Kemp, K.K., Mark, D.M. and Sheppard, E., 1999, Introduction to the Varenius project. *International Journal of Geographical Information Science*, **13**, pp. 731–745.

Goodchild, M.F. and Gopal, S. (Eds), 1989, *Accuracy of Spatial Databases* (Basingstoke: Taylor & Francis).

Goodchild, M.F. and Hunter, G.J., 1997, A simple positional accuracy measure for linear features. *International Journal of Geographical Information Systems*, **11**, pp. 299–306.

Goodchild, M.F. and Lam, N., 1980, Areal interpolation: a variant of the traditional spatial problem. *Geo-Processing*, **1**, pp. 297–312.

Goodchild, M.F. and Moy, W-S, 1976, Estimation from grid data: the map as a stochastic process. In *Proceedings of the Commission on Geographical Data Sensing and Processing*, Moscow, edited by Tomlinson, R.F. (Ottawa: International Geographical Union, Commission on Geographical Data Sensing and Processing, 1977), pp. 67–77.

Goodchild, M.F., Sun, G. and Yang, S., 1992, Development and test of an error model for categorical data. *International Journal of Geographical Information Systems*, **6**, pp. 87–104.

Goovaerts, P., 1996, Stochastic simulation of categorical variables using a classification algorithm and simulated annealing. *Mathematical Geology*, **28**, pp.

909–921.

Gopal, S. and Woodcock, C., 1994, Theory and methods for accuracy assessment of thematic maps using fuzzy sets. *Photogrammetric Engineering and Remote Sensing*, **60**, pp. 181–188.

Gregory, S., 1978, *Statistical Methods and the Geographer*, 4th edition (London: Longman).

Grenander, U., 1996, *Elements of Pattern Theory* (Baltimore, Maryland: The Johns Hopkins University Press).

Greve, C. (Ed.), 1996, *Digital Photogrammetry: an addendum to the Manual of Photogrammetry* (Bethesda, MD: American Society of Photogrammetry and Remote Sensing).

Griffith, D.A. and Layne, L.J., 1999, *A Casebook for Spatial Statistical Data Analysis* (New York: Oxford University Press).

Grün, A., 1982, The accuracy potential of the modern bundle block adjustment in aerial photogrammetry. *Photogrammetric Engineering and Remote Sensing*, **48**, pp. 45–54.

Guptill, S.C. and Morrison, J.L. (Eds), 1995, *Elements of Spatial Data Quality* (New York: Elsevier Science).

Gurney, C.M. and Townshend, J.R.G., 1983, The use of contextual information in the classification of remotely sensed data. *Photogrammetric Engineering and Remote Sensing*, **49**, pp. 55–64.

Haining, R., Griffith, D.A. and Bennett, R., 1983, Simulating two-dimensional autocorrelated surfaces. *Geographical Analysis*, **15**, pp. 247–255.

Hallert, B., 1960, *Photogrammetry, Basic Principles and General Survey* (New York: McGraw Hill).

Harley, J.B., 1975, *Ordnance Survey Maps: a Descriptive Manual* (Southampton: Ordnance Survey).

Harris, R., 1987, *Satellite Remote Sensing: An Introduction* (London and New York: Routledge and Kegan Paul).

Hastings, W.K., 1970, Monte-Carlo sampling methods using Markov chains and their applications. *Biometrika*, **57**, pp. 97–109.

Hay, A.M., 1979, Sampling designs to test land-use map accuracy. *Photogrammetric Engineering and Remote Sensing*, **45**, pp. 529–533.

Helava, U.V., 1978, Digital correlation in photogrammetric instruments. *Photogrammetria*, **34**, pp. 19–41.

Heuvelink, G.B.M., 1998, *Error Propagation in Environmental Modeling with GIS* (London: Taylor & Francis).

Holmes, K.W., Chadwick, O.A. and Kyriakidis, P.C., 2000, Error in a USGS 30-meter digital elevation model and its impact on terrain modeling. *Journal of Hydrology*, **233**, pp. 154–173.

Hord, R.M. and Brooner, W., 1976, Land-use map accuracy criteria. *Photogrammetric Engineering and Remote Sensing*, **42**, pp. 671–677.

Hunter, G.J. and Goodchild, M.F., 1995, Dealing with error in spatial databases: a simple case study. *Photogrammetric Engineering and Remote Sensing*, **61**, pp. 529–537.

Hunter, G.J. and Goodchild, M.F., 1996, A new model for handling vector data uncertainty in Geographic Information Systems. *Journal of the Urban and Regional Information Systems Association*, **8**, pp. 51–57.

Hunter, G.J. and Goodchild, M.F., 1997, Modeling the uncertainty of slope and aspect estimates derived from spatial databases. *Geographical Analysis*, **29**, pp. 35–49.

Huntington, E.V., 1955, *The Continuum and Other Types of Serial Order*, 2nd edition (New York: Dover Publications, Inc.).

Hutchinson, C.F., 1982, Techniques for combining Landsat and ancillary data for digital classification improvement. *Photogrammetric Engineering and Remote Sensing*, **48**, pp. 123–130.

Jager, G. and Benz, U., 2000, Measures of classification accuracy based on fuzzy similarity. *IEEE Transactions on Geoscience and Remote Sensing*, **38**, pp. 1462–1467.

Jain, A.K., Zhong, Y. and Lakshmanan, S., 1996, Object matching using deformable templates. *IEEE Transactions on Pattern Analysis and Machine Intelligence*, **18**, pp. 267–277.

Jeffreys, H., 1983, *Theory of Probability*, 3rd edition (Oxford: Clarendon Press).

Jensen, J.R., 1979, Spectral and textural features to classify elusive land cover at the urban fringe. *The Professional Geographer*, **31**, pp. 400–409.

Jenson, S.K. and Domingue, J.O., 1988, Extracting topographic structure from digital elevation data for geographic information system analysis. *Photogrammetric Engineering and Remote Sensing*, **54**, pp. 1593–1600.

João, E.M., 1998, *Causes and Consequences of Map Generalization* (London: Taylor & Francis).

Johnston, R.J., 1968, Choice in classification: the subjectivity of objective methods. *Annals of the Association of American Geographers*, **58**, pp. 575–589.

Journel, A.G., 1983, Non-parametric estimation of spatial distributions. *Mathematical Geology*, **15**, pp. 445–468.

Journel, A.G., 1986, Geostatistics: models and tools for the Earth sciences. *Mathematical geology*, **18**, pp. 119–140.

Journel, A.G., 1996, Modelling uncertainty and spatial dependence: stochastic imaging. *International Journal of Geographical Information Systems*, **10**, pp. 517–522.

Journel, A. G. and Huijbregts, C.J., 1978, *Mining Geostatistics* (London and New York: Academic Press).

Jupp, D.L.B., Strahler, A.H. and Woodcock, C.E., 1988, Autocorrelation and regularization in digital images. I. Basic theory. *IEEE Transactions on Geoscience and Remote Sensing*, **26**, pp. 463–473.

Kaplan, E.D. (Ed.), 1996, *Understanding GPS: Principles and Applications* (Boston: Artech House).

Keefe, R., 2000, *Theories of Vagueness* (Cambridge: Cambridge University Press).

Keefer, B.J., Smith, J.L. and Gregoire, T.G., 1991, Modeling and evaluating the effects of stream mode digitizing errors on map variables. *Photogrammetric Engineering and Remote Sensing*, **57**, pp. 957–963.

Kemp, K.K., 1997, Fields as a framework for integrating GIS and environment process models. Part One: representing spatial continuity. *Transactions in GIS*, **1**, pp. 219–234, and page 335.

Kiiveri, H.T., 1997, Assessing, representing and transmitting positional uncertainty in maps. *International Journal of Geographical Information Systems*, **11**, pp. 33–52.

Kiiveri, H. T., Caccetta, P. and Evans, F., 2001, Use of conditional probability networks for environmental monitoring. *International Journal of Remote Sensing*, **22**, pp. 1173–1190.

King, P.R. and Smith, P.J., 1988, Generation of correlated properties in heterogeneous porous media. *Mathematical Geology*, **20**, pp. 863–877.

Kirby, R.P., 1988, Measuring the areas of rural land use parcels. In *Rural Information for Forward Planning*, edited by Bunce, R.G.H. and Barr, C.J. (Grange-over-Sands: Institute of Terrestrial Ecology), pp. 99–104.

Kirby, R.P., 1992, The 1987-1989 Scottish national aerial photographic initiative. *Photogrammetric Record*, **14**, pp. 187–200.

Koczkodaj, W.W., Orlowski, M. and Marek, V.W., 1998, Myths about rough set theory. *Communications of the Association for Computing Machine*, **41**, pp. 102–103.

Kolmogorov, A.N., 1956, *Foundations of the Theory of Probability*, 2nd English edition, translation edited by Morrison, N. (New York: Chelsea Publishing Company).

Konecny, G., 1979, Methods and possibilities for digital differential rectification. *Photogrammetric Engineering and Remote Sensing*, **45**, pp. 727–734.

Kraus, K. and Pfeifer, N., 1998, Determination of terrain models in wooded areas with airborne laser scanner data. *ISPRS Journal of Photogrammetry and Remote Sensing*, **53**, pp. 193–203.

Krige, D.G., 1962, Statistical applications in mine valuation. *Journal of the Institute of Mine Surveyors of South Africa*, **12**, pp. 45–84, 95–136.

Kyburg, H.E., Jr. and Smokler, H.E., 1964, *Studies in Subjective Probability* (New York: John Wiley & Sons, Inc.).

Kyriakidis, P.C., Shortridge, A.M. and Goodchild, M.F., 1999, Geostatistics for conflation and accuracy assessment of digital elevation models. *International Journal of Geographical Information Science*, **13**, pp. 677–707.

Lane, S.N., 2000, The measurement of river channel morphology using digital photogrammetry. *Photogrammetric Record*, **16**, pp. 937–961.

Langford, M. and Unwin, D.J., 1994, Generating and mapping population density surface within a geographical information system. *The Cartographic Journal*, **31**, pp. 21–26.

Lark, R.M., 1998, Forming spatially coherent regions by classification of multivariate data: an example from the analysis of maps of crop yield. *International Journal of Geographical Information Science*, **12**, pp. 83–98.

Lark, R.M. and Bolam, H.C., 1997, Uncertainty in prediction and interpretation of spatially variable data on soils. *Geoderma*, **77**, pp. 263–282.

Lee, T., Richards, J.A. and Swain, P.H., 1987, Probabilistic and evidential approaches for multisource data analysis. *IEEE Transactions on Geoscience and Remote Sensing*, GE-**25**, pp. 283–293.

Leung, Y., 1987, On the imprecision of boundaries. *Geographical Analysis*, **19**, pp. 125–151.

Leung, Y. and Yan, J., 1998, A locational error model for spatial features. *International Journal of Geographical Information Science*, **12**, pp. 607–620.

Lewis, M.M., 1998, Numeric classification as an aid to spectral mapping of vegetation communities. *Plant Ecology*, **136**, pp. 133–149.

Li, Z., 1994, A comparative study of the accuracy of digital terrain models (DTMs) based on various data models. *ISPRS Journal of Photogrammetry and Remote Sensing*, **49**, pp. 2–11.

Lillesand, T.M. and Kiefer, R.W., 1994, *Remote Sensing and Image Interpretation*, 3rd edition (New York: John Wiley & Sons, Inc.).

Liu, H. and Jezek, K.C., 1999, Investigating DEM error patterns by directional variograms and Fourier analysis. *Geographical Analysis*, **31**, pp. 249–266.

Longley, P.A., Goodchild, M.F., Maguire, D.J. and Rhind, D.W. (Eds), 1999, *Geographical Information Systems: Principles, Techniques, Applications and Management*, 2nd edition (New York: John Wiley & Sons, Inc.).

Longley, P.A., Goodchild, M.F., Maguire, D.J. and Rhind, D.W., 2001, *Geographic Information Systems and Science* (New York: John Wiley & Sons, Inc.).

Lowell, K.E., 1995, A fuzzy surface cartographic representation for forestry based on Voronoi Diagram area stealing. *Canadian Journal of Forestry Resources*, **24**, pp. 1970–1980.

MacDonald, R.B. and Hall, F.G., 1980, Global crop forecasting. *Science*, **208**, pp. 670–679.

Maling, D.H., 1989, *Measurement from Maps: Principles and Methods of Cartometry* (Oxford: Pergamon).

Mapping Science Committee, National Research Council, 1993, *Towards a Coordinated Spatial Data Infrastructure for the Nation* (Washington, D.C.: National Academy Press).

MacEachren, A.M., 1994, *Some Truth with Maps: A Primer on Symbolization and Design* (Washington, D.C.: Association of American Geographers).

Mark, D.M., 1978, Concepts of data structure for digital terrain models. In *Proceedings Digital Terrain Models (DTM) Symposium*, St Louis, Missouri, pp. 24–31.

Mark, D.M. and Csillag, F., 1989, The nature of boundaries on area-class maps. *Cartographica*, **21**, pp. 65–78.

Maselli, F., Gilabert, M.A. and Conese, C., 1998, Integration of high and low resolution NDVI data for monitoring vegetation in Mediterranean environments. *Remote Sensing of Environment*, **63**, pp. 208–218.

Mather, P.M., 1987, *Computer Processing of Remotely Sensed Images: An Introduction* (Chichester: John Wiley & Sons, Inc.).

Matheron, G., 1963, Principles of geostatistics. *Economic Geology*, **58**, pp. 1246–1266.

McCullagh, M.J., 1988, Terrain and surface modelling systems: theory and practice. *Photogrammetric Record*, **12**, pp. 747–779.

Meek, D.S. and Walton, D.J., 1990, Several methods for representing discrete data by line segments. *Cartographica*, **28**, pp. 13–20.

Metropolis, N., Rosenbluth, A.W., Rosenbluth, M.N., Teller, A.H. and Teller, E., 1953, Equations of state calculations by fast computing machines. *The Journal of Chemical Physics*, **21**, pp. 1087–1092.

Mikhail, E.M., Bethel, J.S. and McGlone, J.C., 2001, *Introduction to Modern Photogrammetry* (New York: John Wiley & Sons, Inc.).

Molenaar, M., 1998, *An Introduction to the Theory of Spatial Object Modelling for GIS* (London: Taylor & Francis).

Moore, I.D., Grayson, R.B. and Ladson, A.R., 1993, Digital terrain modelling: a review of hydrological, geomorphological, and biological applications. In *Terrain Analysis and Distributed Modelling in Hydrology*, edited by Beven, K.J. and Moore, I.D. (Chichester: John Wiley & Sons, Inc.), pp. 7–24.

Moran, P.A.P., 1948, The interpretation of statistical maps. *Journal of the Royal Statistical Society*, Series B, **10**, pp. 243–240.

Morehouse, S., 1985, ARC/INFO: a geo-relational model for spatial information. In *Proceedings Auto-Carto 7, the Seventh International Symposium on Computer-Assisted Cartography*, Washington, D.C., pp. 388–397.

Müller, J.C., 1977, Map gridding and cartographic errors: a recurrent argument. *The Canadian Cartographer*, **14**, pp. 152–167.

Myers, D.E., 1982, Matrix formulation of Cokriging. *Mathematical Geology*, **14**, pp. 249–257.

Myers, D.E., 1991, Pseudo-cross variograms, positive-definiteness, and Cokriging. *Mathematical Geology*, **23**, pp. 805–816.

National Committee for Digital Cartographic Data Standards (NCDCDS), 1988, The proposed standard for digital cartographic data. *The American Cartographer*, **15**, pp. 9–140.

Newcomer, J.A. and Szajgin, J., 1984, Accumulation of thematic map errors in digital overlay analysis. *The American Cartographer*, **11**, pp. 58–62.

Nguyen, H.T., 1997, Fuzzy sets and probability. *Fuzzy Sets and Systems*, **90**, pp. 129–132.

Noronha, V.T., 2000, Towards ITS map database interoperability – database error and rectification. *GeoInformatica*, **4**, pp. 201–213.

Noronha, V.T., Goodchild, M.F., Church, R.L. and Fohl, P., 1999, Location expression standards for ITS: testing the LRMS cross street profile. *Annals of Regional Science*, **33**, pp. 197–212.

Oliver, M.A. and Webster, R., 1989, A geostatistical basis for spatial weighting in multivariate classification. *Mathematical Geology*, **21**, pp. 15–35.

Oliver, M.A. and Webster, R., 1990, Kriging: a method of interpolation for geographical information systems. *International Journal of geographical Information Systems*, **4**, pp. 313–332.

Openshaw, S., 1977, A geographical solution to scale and aggregation problems in region-building, partitioning and spatial modelling. *Transactions of the Institute of British Geographers*, **2**, pp. 459–472.

Pawlak, Z., 1982, Rough sets. *International Journal of Computer and Information Sciences*, **11**, pp. 341–356.

Pawlak, Z., 1998, Rough set theory and its applications to data analysis. *Cybernetics and Systems: An International Journal*, **29**, pp. 661–688.

Pawlowsky, V., Olea, R.A. and Davis, J.C., 1995, Estimation of regionalized compositions: a comparison of three methods. *Mathematical Geology*, **27**, pp. 105–127.

Perkal, J., 1956, On epsilon length. *Bulletin de l'Academie Polonaise des Sciences*, **4**, pp. 399–403.

Peucker, T.K. and Douglas, D.H., 1975, Detection of surface specific points by local parallel processing of discrete terrain elevation data. *Computer Graphics and Image Processing*, **4**, pp. 375–387.

Peuquet, D.J., 1984, A conceptual framework and comparison of spatial data models. *Cartographica*, **21**, pp. 66–113.

Peuquet, D.J., 1988, Representations of geographic space: toward a conceptual synthesis. *Annals of the Association of American Geographers*, **18**, pp. 375–394.

Pipkin, J.S., 1978, Fuzzy sets and spatial choice. *Annals of the Association of American Geographers*, **68**, pp. 196–204.

Prisley, S.P., Gregoire, T.G. and Smith, J.L., 1989, The mean and variance of area estimates computed in an arc-node geographical information system. *Photogrammetric Engineering and Remote Sensing*, **55**, pp. 1601–1612.

Quackenbush, L.J., Hopkins, P.F. and Kinn, G.J., 2000, Developing forestry products from high resolution digital aerial imagery. *Photogrammetric Engineering and Remote Sensing*, **66**, pp. 1337–1348.

Rathbun, S.L., 1998, Spatial modelling in irregularly shaped regions: Kriging estuaries. *Environmetrics*, **9**, pp. 109–129.

Richards, J.A., 1996, Classifier performance and map accuracy. *Remote Sensing of Environment*, **57**, pp. 161–166.

Ripley, B.D., 1992, Stochastic models for the distribution of rock types in petroleum reservoirs. In *Statistics in the Environmental and Earth Sciences*, edited by Walden, A.T. and Guttorp, P. (London: Edward Arnold), pp. 247–282.

Robinove, C.J., 1981, The logic of multispectral classification and mapping of land. *Remote Sensing of Environment*, **11**, pp. 231–244.

Rosenberg, P., 1971, Resolution, detectability, and recognizability. *Photogrammetric Engineering and Remote Sensing*, **37**, pp. 1244–1258.

Rosenfield, G.H. and Melley, M.L., 1980, Applications of statistics to thematic mapping. *Photogrammetric Engineering and Remote Sensing*, **46**, pp. 1287–1294.

Rossi, R.E., Dungan, J.L. and Beck, L.R., 1994, Kriging in the shadows: geostatistical interpolation for remote sensing. *Remote Sensing of Environment*, **49**, pp. 32–40.

Roy, D.P., Kennedy, P. and Folying, S., 1997, Combination of the normalised difference vegetation index and surface temperature for regional scale European forest cover mapping using AVHRR data. *International Journal of Remote*

Sensing, **18**, pp. 1189–1195.

Ruspini, E.H., 1969, A new approach to clustering. *Information and Control*, **15**, pp. 22–32.

Saalfeld, A., 1988, Conflation: automated map compilation. *International Journal of Geographical Information Systems*, **2**, pp. 217–228.

Saaty, T.L., 1978, Exploring the interface between hierarchies, multiple objectives and fuzzy sets. *Fuzzy Sets and Systems*, **1**, pp. 57–68.

Schenk, T., 1994, Concepts and algorithms in digital photogrammetry. *ISPRS Journal of Photogrammetry and Remote Sensing*, **49**, pp. 2–8.

Schowengerdt, R.A., 1997, Remote Sensing, Models and Methods for Image Processing, 2nd edition (San Diego: Academic Press).

Schreier, H., Lougheed, J., Gibson, J.R. and Russell, J., 1984, Calibrating an airborne laser profiling system. *Photogrammetric Engineering and Remote Sensing*, **57**, pp. 965–971.

Shafer, G., 1986, The combination of evidence. *International Journal of Intelligent Systems*, **1**, pp. 155–179.

Shi, W., 1998, A generic statistical approach for modelling errors of geometric features in GIS. *International Journal of Geographical Information Science*, **12**, pp. 131–143.

Sircar, J.K. and Cebrian, J.A., 1991, An automated approach for labeling raster digitized contour maps. *Photogrammetric Engineering and Remote Sensing*, **57**, pp. 965–971.

Skidmore, A.K. and Turner, B.J., 1988, Forest mapping accuracies are improved using a supervised nonparametric classifier with SPOT data. *Photogrammetric Engineering and Remote Sensing*, **54**, pp. 1415–1421.

Skidmore, A.K. and Turner, B.J., 1992, Map accuracy assessment using line intersect sampling. *Photogrammetric Engineering and Remote Sensing*, **58**, pp. 1453–1457.

Smith, C.A.B., 1961, Consistency in statistical inference and decision. *Journal of the Royal Statistical Society*, series B, **23**, pp. 1–37.

Solow, A.R., 1986, Mapping by simple indicator Kriging. *Mathematical Geology*, **18**, pp. 335–352.

Song, C., Woodcock, C.E., Seto, K.C., Lenney, M.P. and Macomber, S.A., 2001, Classification and change detection using Landsat TM data: when and how to correct atmospheric effects? *Remote Sensing of Environment*, **75**, pp. 230–244.

Spitzer, F., 1971, Markov random fields and Gibbs ensembles. *The American Mathematical Monthly*, **78**, pp. 142–154.

Srinivasan, A. and Richards, J.A., 1990, Knowledge-based techniques for multi-source classification. *International Journal of Remote Sensing*, **11**, pp. 505–525.

Steele, B.M., Winne, J.C. and Redmond, R.L., 1998, Estimation and mapping of misclassification probabilities for thematic land cover maps. *Remote Sensing of Environment*, **66**, pp. 192–202.

Stehman, S.V. and Czaplewski, R.L., 1998, Design and analysis for thematic map accuracy assessment: fundamental principles. *Remote Sensing of Environment*, **64**, pp. 331–344.

Steven, M.D., 1987, Ground truth: an underview. *International Journal of Remote Sensing*, **8**, pp. 1033–1038.

Stevens, S.S., 1946, On the theory of scales of measurement. *Science*, **103**, pp. 677–680.

Strahler, A.H., 1980, The use of prior probabilities in maximum likelihood classification of remotely sensed data. *Remote Sensing of Environment*, **10**, pp. 135–163.

Sullivan, J., 1984, Conditional recovery estimation through probability Kriging – theory and practice. In *Geostatistics for Natural Resources Characterization* (Part I), edited by Verly, G., David, M., Journel, A.G. and Marechal, A. (Amsterdam: D. Reidel Publishing Company), pp. 365–384.

Swain, P.H. and Davis, S.M. (Eds), 1978, *Remote Sensing: the Quantitative Approach* (London and New York: McGraw-Hill International Book Co.).

Switzer, P., 1975, Estimation of accuracy of qualitative maps. In *Display and Analysis of Spatial Data*, edited by Davis, J.C. and McCullagh, M.J. (London and New York: John Wiley & Sons, Inc.), pp. 1–13.

Switzer, P., 1980, Extension of discriminant analysis for statistical classification of remotely sensed satellite imagery. *Mathematical Geology*, **12**, pp. 367–376.

Thapa, K. and Burtch, R.C., 1991, Primary and secondary methods of data collection in GIS/LIS. *Surveying and Land Information Systems*, **51**, pp. 162–170.

Theobald, D.M., 2000, Reducing linear and perimeter measurement errors in raster-based data. *Cartography and Geographic Information Science*, **27**, pp. 111–116.

Thiessen, A.H., 1911, Precipitation for large areas. *Monthly Weather Review*, **39**, pp. 1082–1084.

Thomas, I.L. and Allcock, M.G., 1984, Determining the confidence level for a classification. *Photogrammetric Engineering and Remote Sensing*, **50**, pp. 1491–1496.

Tobler, W.R., 1970, A computer movie simulating urban growth in the Detroit region. *Economic Geography*, (supplement), **46**, pp. 234–240.

Tobler, W.R., 1988, Resolution, resampling, and all that. In *Building Databases for Global Science*, edited by Mounsey, H. and Tomlinson, R. (London, Taylor & Francis), pp. 129–137.

Torlegård, A.K.I., 1986, Some photogrammetric experiments with digital image processing. *Photogrammetric Record*, **12**, pp. 175–196.

Torlegård, A.K.I., Östaman, A. and Lindgren, R., 1986, A comparative test of photogrammetrically sampled digital elevation models. *Photogrammetria*, **41**, pp. 1–16.

Toutin, T. and Gray, L., 2000, State-of-the-art of elevation extraction from satellite SAR data. *ISPRS Journal of Photogrammetry and Remote Sensing*, **55**, pp. 13–33.

Townsend, F.E., 1986, The enhancement of computer classifications by logical smoothing. *Photogrammetric Engineering and Remote Sensing*, **52**, pp. 213–221.

Townshend, J.R.G., 1981, *Terrain Analysis and Remote Sensing* (London: George Allen and Unwin).

Townshend, J.R.G. and Justice, C., 1981, Information extraction from remotely sensed data. A user view. *International Journal of Remote Sensing*, **2**, pp. 313–329.

Townshend, P.A., 2000, A quantitative fuzzy approach to assess mapped vegetation classifications for ecological applications. *Remote Sensing of Environment*, **72**, pp. 253–267.

Trinder, J.C., 1973, Pointing precision, spread functions and MTF. *Photogrammetric Engineering*, **39**, pp. 863–874.

Trinder, J.C., 1989, Precision of digital target location. *Photogrammetric Engineering and Remote Sensing*, **55**, pp. 883–886.

Tudor, G.S. and Sugarbaker, L.J., 1993, GIS orthographic digitizing of aerial photographs by terrain modelling. *Photogrammetric Engineering and Remote Sensing*, **59**, pp. 499–503.

Turner, D.P., Cohen, W.B., Kennedy, R.E., Fassnacht, K.S. and Briggs, J.M., 1999, Relationships between leaf area index and Landsat TM spectral vegetation indices across three temperate zone sites. *Remote Sensing of Environment*, **70**, pp. 52–68.

Tyler, D.A., 1987, Position tolerance in land surveying. *Journal of Surveying Engineering*, **113**, pp. 152–163.

Tveite, H. and Langaas, S., 1999, An accuracy assessment method for geographical line data sets based on buffering. *International Journal of Geographical Information Science*, **13**, pp. 27–47.

U.S. Geological Survey (USGS), 1997, *Standards for Digital Elevation Models*, *National Mapping Program Technical Instructions*. Department of the Interior, USGS, Reston, Virginia.

University Consortium for Geographic Information Science (UCGIS), 1996, Research priorities for geographic information science. *Cartography and Geographic Information Science*, **23**, pp. 115–127.

Vckovski, A., 1998, *Interoperable and Distributed Processing in GIS* (London: Taylor & Francis).

Veregin, H., 1989, Error modelling for map overlay. In *Accuracy of Spatial Databases*, edited by Goodchild, M.F. and Gopal, S. (Basingstoke: Taylor & Francis), pp. 3–18.

Veregin, H., 1995, Developing and testing of an error propagation model for GIS overlay operations. *International Journal of Geographical Information Systems*, **9**, pp. 595–619.

Viertl, R., 1997, On statistical inference for non-precise data. *Environmetrics*, **18**, pp. 541–568.

Vonderohe, A.P., Fisher, S.S. and Krohn, D.K., 1990, A positional comparison of the results of two digital cadastral mapping methods. *Surveying and Land Information Systems*, **50**, pp. 11–24.

Walsh, S.J., Lightfoot, D.R. and Butler, D.R., 1987, Recognition and assessment of error in GIS. *Photogrammetric Engineering and Remote Sensing*, **53**, pp.

1423–1430.

Wang, F., 1990, Improving remote sensing image analysis through fuzzy information representation. *Photogrammetric Engineering and Remote Sensing*, **56**, pp. 1163–1169.

Wang, F. and Hall, G.B., 1996, Fuzzy representation of geographical boundaries in GIS. *International Journal of Geographical Information Systems*, **9**, pp. 573–590.

Wang, Z., 1990, *Principles of Photogrammetry (with Remote Sensing)*, (Beijing: Press of Surveying and Mapping).

Weber, C., 1994, Per-zone classification of urban land cover for urban population estimation. In *Environmental Remote Sensing from Regional to Global Scales*, edited by Foody, G.M. and Curran, P. (Chichester: John Wiley & Sons, Inc.), pp. 142–149.

Webster, R. and Beckett, P.H.T., 1968, Quality and usefulness of soil maps. *Nature*, **219**, pp. 680–682.

Wolf, P.R. and Ghilani, C.D., 1997, *Adjustment Computations: Statistics and Least Squares in Surveying and GIS*, 3rd edition (New York: John Wiley & Sons, Inc.).

Woodcock, C. and Gopal, S., 2000, Fuzzy set theory and thematic maps: accuracy assessment and area estimation. *International Journal of Geographical Information Science*, **14**, pp. 153–172.

Woodcock, C., Gopal, S. and Albert, W., 1996, Evaluation of the potential for providing secondary labels in vegetation maps. *Photogrammetric Engineering and Remote Sensing*, **62**, pp. 393–399.

Woodcock, C. and Harward, V.J., 1992, Nested-hierarchical scene models and image segmentation. *International Journal of Remote Sensing*, **13**, pp. 3167–3187.

Worboys, M., 1998, Computation with imprecise geospatial data. *Computers, Environment and Urban Systems*, **22**, pp. 85–106.

Wright, D.J., Goodchild, M.F. and Proctor, J.D., 1997, Demystifying the persistent ambiguity of GIS as 'tool' versus 'science'. *Annals of the Association of American Geographers*, **87**, pp. 346–362.

Wrobel, B.P., 1991, The evolution of digital photogrammetry from analytical photogrammetry. *Photogrammetric Record*, **13**, pp. 765–776.

Yager, R.R., 1991, On probabilities induced by multi-valued mappings. *Fuzzy Sets and Systems*, **42**, pp. 301–314.

Zadeh, L.A., 1965, Fuzzy sets. *Information and Control*, **8**, pp. 335–353.

Zadeh, L.A., 1968, Probability measures of fuzzy events. *Journal of Mathematical Analysis and Applications*, **23**, pp. 421–427.

Zadeh, L.A., 1978, Fuzzy sets as a basis for a theory of possibility. *Fuzzy Sets and Systems*, **1**, pp. 3–28.

Zadeh, L.A., 1984, Fuzzy probabilities. *Information Processing and Management*, **20**, pp. 363–372.

Zadeh, L.A., 1997, Towards a theory of fuzzy information granulation and its centrality in human reasoning and fuzzy logic. *Fuzzy Sets and Systems*, **90**, pp.

111–127.

Zarkovich, S.S., 1966, *Quality of Statistical Data* (Rome: Food and Agriculture Organization of the United Nations).

Zhang, J., 1996, *A surface-based approach to the handling of uncertainties in an urban-orientated spatial database.* Unpublished PhD thesis, The University of Edinburgh, Scotland.

Zhang, J. and Foody, G.M., 1998, A fuzzy classification of sub-urban land cover from remotely sensed imagery. *International Journal of Remote Sensing*, **19**, pp. 2721–2738.

Zhang, J. and Foody, G.M., 2001, Fully-fuzzy supervised classification of sub-urban land cover from remotely sensed imagery: statistical and artificial neural network approaches. *International Journal of Remote Sensing*, **22**, pp. 615–628.

Zhang, J. and Kirby, R.P., 1997, An evaluation of fuzzy approaches to mapping land cover from aerial photographs. *ISPRS Journal of Photogrammetry and Remote Sensing*, **52**, pp. 193–201.

Zhang, J. and Kirby, R.P., 1999, Alternative criteria for defining fuzzy boundaries based on fuzzy classification of aerial photographs and satellite images. *Photogrammetric Engineering and Remote Sensing*, **65**, pp. 1379–1387.

Zhang, J. and Kirby, R.P. 2000, A geostatistical approach to modelling positional errors in vector data. *Transactions in GIS*, **4**, pp. 145–159.

Zhang, J. and Stuart, N., 2001, Fuzzy methods for categorical mapping with image-based land cover data. *International Journal of Geographical Information Science*, **15**, pp. 175–195.

Zhang, Q., 1998, Fuzziness-vagueness-generality-ambiguity. *Journal of Pragmatics*, **29**, pp. 13–31.

Zhang, Z., Zhang, J., Liao, M. and Zhang, L., 2000, Automatic registration of multi-source imagery based on global image matching. *Photogrammetric Engineering and Remote Sensing*, **66**, pp. 625–632.

Zhu, A., 2000, Mapping soil landscape as spatial continua; the neural network approach. *Water Resources Research*, **36**, pp. 663–677.

Zhu, H. and Journel, A.G., 1991, Mixture of populations. *Mathematical Geology*, **23**, pp. 647–671.

Index